Fundamentals of Wastewater Treatment

Fundamentals of Wastewater Treatment

Tom Dixon

MURPHY & MOORE
www.murphy-moorepublishing.com

Published by Murphy & Moore Publishing,
1 Rockefeller Plaza,
New York City, NY 10020, USA
www.murphy-moorepublishing.com

Fundamentals of Wastewater Treatment
Tom Dixon

International Standard Book Number: 978-1-63987-251-0 (Hardback)

Cataloging-in-Publication Data

Fundamentals of wastewater treatment / Tom Dixon.
p. cm.
Includes bibliographical references and index.
ISBN 978-1-63987-251-0
1. Sewage--Purification. 2. Sewage--Purification--Environmental aspects.
3. Sewage disposal plants. I. Dixon, Tom.
TD745 .F86 2022
628.3--dc23

TABLE OF CONTENTS

PREFACE

Water which has been contaminated due to human use is termed as wastewater. It is the byproduct of various human activities and, can be classified into industrial wastewater, municipal wastewater and domestic wastewater depending upon its source. The techniques and methods which are used to remove contaminants and other unwanted particles from water fall under the discipline of wastewater treatment. Some of the common techniques used in a wastewater treatment facility are sedimentation, biochemical oxidation, polishing and chemical oxidation. According to the water being treated, treatment plants can be classified into industrial wastewater treatment plants, sewage treatment plants, leachate treatment plants and agricultural wastewater treatment plants. This book is a compilation of chapters that discuss the most vital concepts in the field of wastewater treatment. It presents selected concepts that redefine wastewater treatment. This book is an essential guide for both academicians and those who wish to pursue this discipline further.

To facilitate a deeper understanding of the contents of this book a short introduction of every chapter is written below:

Chapter 1- Wastewater is any water that has been used and has been affected by domestic, industrial and commercial use. Wastewater is segregated into three distinctive groups based on the degree of contamination of water– black water, grey water and yellow water. This chapter serves as an introduction to wastewater, its types and its sources.

Chapter 2- Wastewater management is a combination of one or more ways through which wastewater is collected, stored, treated and finally disposed of. It is intrinsically related to the management of the water chain. Wastewater quality and quality indicators are two important features of wastewater management. This topic has been carefully written to provide a better understanding of the topic.

Chapter 3- Wastewater pollution refers to the contamination of water bodies, land and ground water by wastewater discharge. Wastewater contains many harmful chemicals and substances that render water unfit for consumption and other purposes, and even affects the soil. This chapter sheds light on wastewater pollution and its effects on the different systems.

Chapter 4- A sewer is a large, artificial underground passage or pipe that carries waste and used water away for treatment. Sewer system design plays an important role in city planning because it decides the method of efficient wastewater disposal. Sewer systems are provided with openings, called manholes, through which inspection, maintenance and upgrades can be done. This chapter discusses in detail the different aspects of sewers and manholes.

Chapter 5- Sewage is the waste matter or liquid refuse carried off by sewers. They pose serious health and environmental risks if not handled and properly disposed. This chapter takes a closer look at sanitation processes like primary, secondary and tertiary treatments, the different sewage systems and sewage handling methods in order to provide a clear understanding of the subject matter.

Chapter 6- Wastewater treatment is defined as the removal of impurities and contaminants from wastewater or sewage before it is released back into the environment. Wastewater can be treated through a number of physical, chemical and biological processes or any combination of the three. Sedimentation wastewater treatment, sewage sludge treatment, bar screen, etc. are some techniques which are employed for the treatment of wastewater. This chapter is a comprehensive summary of these wastewater treatment techniques.

I would like to share the credit of this book with my editorial team who worked tirelessly on this book. I owe the completion of this book to the never-ending support of my family, who supported me throughout the project.

Tom Dixon

Chapter 1

Understanding Wastewater

Wastewater is any water that has been used and has been affected by domestic, industrial and commercial use. Wastewater is segregated into three distinctive groups based on the degree of contamination of water– black water, grey water and yellow water. This chapter serves as an introduction to wastewater, its types and its sources.

Wastewater is affected by domestic, industrial and commercial use, thus constantly changing its composition and making it rather difficult to define.

Wastewater is used water that has been affected by domestic, industrial and commercial use. The composition of all wastewaters is thus constantly changing and highly variable, which is why it is so difficult to pinpoint a singular definition of the word itself.

The composition of wastewater is 99.9% water and the remaining 0.1% is what is removed. This 0.1%

contains organic matter, microorganisms and inorganic compounds. Wastewater effluents are released to a variety of environments, such as lakes, ponds, streams, rivers, estuaries and oceans. Wastewater also includes storm runoff, as harmful substances wash off roads, parking lots and rooftops.

Types of Wastewater

Often used interchangeably with the term sewage, "sewage" technically denotes any wastewaters which pass through a sewer. Prior to entering a wastewater treatment plant, wastewater is sometimes called raw wastewater or raw sewage.

Domestic wastewater originates from activities such as restroom usage, bathing, food preparation and laundry. Commercial wastewater from non-domestic sources, such as beauty salons or auto body repair shops, for example. This wastewater may contain hazardous materials and requires special treatment or disposal. Industrial wastewater originates from industrial or commercial manufacturing processes, such as agriculture, and are usually more difficult to treat than domestic wastes. Industrial wastewater's composition varies on an industry-by-industry basis.

Organic Content of Wastewater

The organic content of wastewater is made up of human feces, protein, fat, vegetable and sugar material from food preparation, as well as soaps. Some of this organic content is dissolved into the water and some exist as separate particles. The portion of organic material that does not dissolve but remains suspended in the water is known as suspended solids. Wastewater is treated to remove as much organic material as possible.

Implications for Microorganisms

Naturally occurring soil and water bacteria eat the organic waste in wastewater and use it as a food and energy source to grow rapidly. In a natural water environment where there is plenty of oxygen dissolved in the water, aerobic bacteria eat the organic material and form a slime of new bacterial cells and dissolved salt-waste products.

If undiluted wastewater is left on its own, anaerobic bacteria decompose the waste organic material and release odorous gases such as hydrogen sulphide. Odor-free gases such as methane and carbon dioxide can also be released.

Where there is an overwhelming amount of wastewater, all the oxygen will be used up and the anaerobic bacteria will take over, making the water go septic. This is ultimately harmful to fish and other forms of life dependent on oxygen, on occasion creating dead zones.

Inorganic Matter

Inorganic minerals, metals and compounds, such as sodium, copper, lead and zinc are common in wastewater from both sewage and wastewater. They can originate from industrial and commercial sources, stormwater, and inflow and infiltration from cracked pipes. Most inorganic substances are stable and cannot be broken down easily by organisms in wastewater.

Nutrients

Excessive nutrients such as phosphorus and nitrogen can cause eutrophication, which can also be toxic to aquatic organisms. This also promotes excessive plant growth and reduces oxygen availability, altering habitats and potentially endangering certain species.

Other Wastewater Pollutants

Bacteria, viruses and disease-causing pathogens in wastewater can pollute beaches and contaminate shellfish populations. Fecal coliform bacteria in human waste is typically harmless, but there are pathogens that can negatively impact human health. These can be bacteria such as typhoid or viruses such as hepatitis B, for example. Direct contact with these pathogens or pollution of the water supply can cause infections as a result.

Types of Wastewater

Black Water

Black water is waste water loaded with biological material such as grease or feces. It is colloquially referred to as "sewage" and contains highly dangerous materials that can pose real health hazards to you and your family.

However, some people are deeply unaware of how easy it is to come into contact with this dangerous water type. One of the most obvious sources of black water is the toilet. Human waste, fecal matter, is generally regarded as a biohazard, especially when people are sick: it can contain bacteria and pathogens that could potentially be passed on to others with improper handling. Kitchen sinks can also be a source of black water as they contain grease, oils, and food chunks that can decay and carry germs with them. Even laundry water from a sick person's house can be considered black water, for reasons already stated above: airborne bacteria and pathogens.

Black water can also contain a large amount of pharmaceuticals, especially if it is coming from a home in which a sick person resides. This water should never ever be released into the environment, as its toxic makeup can harm the ecosystem.

Algae is a phytoplankton that feeds directly on fertilizer and reacts the way plants do when exposed to that fertilizer: it grows rapidly, creating algal blooms. Not all of them are harmful, but some are deeply detrimental to one's health and can cause diarrhea, memory loss, nausea and vomiting, and liver failure. Toxic algal blooms can be found in black water.

Black water should be left strictly for a professional to handle in the Indianapolis area: it is too dangerous and toxic for the average homeowner and homemaker to tackle by themselves. Specialized equipment is required in order to clear an area of black water and black water damage safely without causing harm to the health of both the humans and animals who reside in the space. Extended exposure to black water can lead to infections, illnesses, and long-term respiratory problems, while coming into contact with it exposes a human to even more serious health risks and even death in certain cases. The money you think you might be saving is clearly not worth the grave danger to your health.

Greywater

Greywater is described as gently used waste water that comes from sources such as the bathroom, kitchen and laundry. It is relatively clean and cannot be compared with the water from toilets (blackwater). Generally, any water that is drained from the house other than toilet water can be described as greywater. It may also contain traced of food, grease, hair, dirt or particular household cleaning products.

As much as it may look dirty, greywater can be reused for other purposes and does not necessarily have to be disposed of into the sewage system as other types of waste water do. This means that it is not really waste water but could be repurposed for safe and beneficial uses such as yard irrigation. It therefore serves to substitute fresh water for some purposes thus allowing for water conservation as well as reducing the amount of water that is sent into the sewage system for treatment.

Different Ways to Collect Greywater

Manual Collection

Greywater is collected from laundry, kitchen or bath water. The systems for collecting greywater vary from manual to completely automated with the manual systems being the simplest to use and the ones with the least costs of maintenance. The manual systems require nothing more than a bucket and jug for collecting greywater by hand. The water can then be used either for irrigation plants or as a substitute for using fresh water for flushing.

Simple Piping

The second more complex system creates an automated connection between laundry and the landscape. This system uses plumbing connections between washing machine's waste water drain pipe and the lawn. Here, the water flows via underground piping into the yard where it is used to irrigate the land. It has a low cost of materials and requires some maintenance to ensure that the plumbing does not become clogged up with residue from the soapy water. This kind of system can be built by oneself with little effort and training plus the materials required such as hose pipes and connectors.

Integrated Sewage and Piping

The third system involves integrating the whole houses plumbing into a greywater collection system. These systems are the most expensive and the most maintenance intensive options available. They also require that one hires professionals in order to ensure that the work is done properly. The systems usually connect bath, laundry and kitchen drains to a system that dispenses greywater into a collection tank. The water usually needs to be used within 24 hours of collection in order to avoid bacteria from growing to numbers that cause foul smells. The water can however be treated with chlorine or iodine to allow for longer storage.

Some systems can even allow for the merging of greywater to toilet plumbing ensuring that one does not flush the toilet with fresh water. The systems can be further integrated to include greywater outlets that merge into the existing sewage systems to ensure there is no overflow. Some systems can also come with an emergency shut of switch to ensure that in case the system fails, there is no backflow into the main freshwater piping. The amount of piping and pumping involved in these most complex systems means they are more susceptible to blocking due to residual soap. They therefore need more maintenance than other systems.

Benefits of using Greywater

Fresh Water Conservation

Use of greywater reduces the amount of fresh water used up in the household for other functions. The reusing of greywater to flush toilets and water plants reduces the load of fresh water required and if enough people make the same choice, the water demands can be reduced so drastically as to positively impact the environment. This is particularly important in regions experiencing dry climates or going through droughts.

Reduction of Water Wastage

The use of freshwater for all the different functions in a house produce lots of waste water that all ends up flowing into the sewer systems. This creates unnecessary wastage. By reusing greywater for one other function before it is dumped into the sewage, the amount of water wasted can be halved. Thus, it means fewer resources can be diverted to sewage treatment and that the saving of even an extra liter of water which could have been used before its disposal increases as well, thus increasing water use efficiency.

Reduced Energy Consumption

With reduced fresh water demands for each household, there is reduced energy demand required to pump the water into the house. Secondly, water reuse reduces the load that treatment centers have to handle in ensuring that it is purified thus reducing total energy required by both water distribution and sewage purifying companies. Reduced energy use means less electricity and in turn, fossil fuel use is also cut down eventually helping to reduce greenhouse gas emissions. It means that energy can be diverted into other resource intensive requirements or stored for later use.

Reduced Chemical Consumption

The reduction of water sent to sewage plants also leads to reduced levels in the amount of chemicals used in treating the water. With a smaller load, sewage treatment plants do not require as much use of chemicals which is beneficial to the environment as it reduces both the cost of sewage treatment as well as reducing the demand for chemical production that goes to benefit the environment. It also reduces the chance of accidental dumping of chemical waste by sewage processing companies. Greywater use also reduces the need to fertilize the yards with chemical fertilizers due to the nutrients the water already contains.

Beautification of Landscapes

The use of greywater to water yards provides an alternative source of water to plants, especially in regions where plants lack it. Regions that use reclaimed greywater thus provide alternative use for water that would have gone to waste. Therefore, it creates beautiful landscapes as plants maintain greenness and bloom, even in regions where the climates are harsh. Coupled with other technologies such as the use of mulch and drip irrigation, the water provided can last longer to ensure the beautiful state of one's yard is maintained even during the driest of seasons.

Reuse of Nutrients

The use of greywater reintroduces nutrients that would have otherwise been lost to the sewage system. Kitchen and bath water contain a lot of organic material that is not harmful to plants. This is unlike blackwater where the organic levels are too high. Blackwater also has lots of harmful bacteria e.g. E. coli that could cause disease if they contaminate the soil. The greywater provides plants with good nutrients that go towards increased beautification.

Organic Filtering of Water

The greywater used in watering plants is purified by both plant and soil action which negates the use of chemicals that might, in excess, harm the soil and the organisms that live within it. Organic filtering serves to ensure the environments safety.

Groundwater Recharge

The greywater used often ends up being filtered by the soil as it moves down and ends up in the ground water where is replenishes the levels therein. This is a good thing as it ensures waterways such as rivers sourcing their water from the underground aquifers are always fed.

How the Greywater System Works?

The process involves three steps which include:

Collection of Greywater

The greywater can be collected from the sinks, baths or washing machines by use of basins or through piping. The problem that occurs with this is the kinds of soaps used. Some soap have harsh chemical that can damage the plumbing system over time. It is recommended that one uses organic soaps as they are less likely to cause harm.

Secondly, the use of plumbing is recommended as it requires less maintenance and will not tire the individual as much as manual bucket collection will. The system can also make use of pumps to transport the water to collection sites. This allows for storage of waste water while it is not in use. The added challenge from this is the fact that the pumps may be affected by the chemicals within greywater over time.

Storage and Treatment

The water is usually collected in a storage container afterwards. Storage is recommended for less than 24 hours to ensure bad smells do not accumulate. It is also recommended that one treats the water with chlorine or iodine to ensure that the stored water does not produce foul smell, if the plan is to store the water for longer.

Use of Greywater

The greywater system should be connected in the most appropriate way to ensure look term use. It is best to have the system both flush toilets and pour into the yards. The water can be transported via use of gravity or pumps, but ideally, there should be a way to divert water into the usual sewage path if it is not needed. Also, there should be an emergency cut off in case of failure of the system to prevent backflow of the greywater into fresh water systems. The greywater can also find use in any other non-contact way such as with washing garden tools and the driveway. When used for gardening, the plants should not be acidophilic (plants that do well in acidic soils) as greywater is usually alkaline in nature due to the use of soaps.

Yellow Water

Yellow water is essentially pure urine. It's urine from specific sources which doesn't have any of the contaminants found in greywater or blackwater like chemicals, toilet paper, faeces and food particles.

Separate collection of yellow water is possible with sorting toilet. Among the flows of wastewater, yellow water contains most of the nutrients. One person produces on average 3.92 kg of nitrogen, 0.38 kg of phosphorous and 0.97 kg of potassium per year. These nutrients, such as nitrogen in the form of urea, phosphorus as super phosphate and potassium as an ion, are in a form which are ideal for uptake by plant. Beneficially, urine contains very low levels of heavy metals and pathogens. These heavy metal concentrations are much lower than those of most chemical fertiliser. In Sweden, for instance, urine contains less than 3.2 mg cadmium per kg of phosphorus compared to 26 mg Cd/kg of phosphorus in commercial fertiliser and 55 mg Cd/kg of phosphorous in sludge.

In the conventional wastewater treatment systems, instead of utilising the yellow water for plant nutrition, it is wasted. In modern municipal wastewater treatment plants, nitrogen compounds (most of them originating from yellow water) is removed with the costly nitrification and denitrification process. Even with the high inputs of money, a reasonable amount of nitrogen compounds (especially nitrate) escapes with the effluents of treatment plants and causes eutrophication in water bodies. In high-tech treatment plants, most of the nitrogen compounds are converted to N_2, which is itself a raw material for high energy consuming nitrogen fertilizer synthesis (e.g. in natural-gas based ammonia plants).

Even the best affordable treatment plants discharge over 20 % of nitrogen, over 5 % of phosphorus and more than 90 % of potassium to the aquatic environment where they are lost for ever and cause severe problems. Those nutrients, which are captured in sludge are often contaminated with heavy metals such as Cadmium (Cd) and organic compounds such as PCB (polychlorinated Biphenyle), which pose potential toxic risks to plants, animals and humans. Therefore, large amounts of Sewage sludge are disposed of in landfills or incinerated. Only a smaller part is applied to agricultural land. Due to following reasons wasting of yellow water is not sustainable:

- Production of fertiliser is energy intensive, draws on very limited fossil resources and causes environmental problems. Generating nitrogen (N) from air requires a considerable amount of energy. Mining and refining the raw materials for phosphate production generates a huge amount of hazardous wastes. Reserves of phosphate (P) and potassium (K) are definitely limited on a time scale of only a couple of human generations especially with regard to economic constraints. Moreover, also sulphur is a limited fossil resource and is required by plants. Also sulphur is contained in yellow water. Therefore, we should use the human resource "Anthropogenic Nutrient Solution" yellow water as fertiliser which contains reasonable amounts of these nutrients.
- If yellow water is wasted (i.e. it is added to municipal wastewater), it will either contribute to eutrophication of surface waters because of its nutrient content or it will even lead to groundwater contamination (high nitrate concentrations can be generated by transformation of ammonia, and sometimes high nitrate concentrations are detected in ground waters contaminated with domestic wastewater - they can lead to lethal methemoglobinemia of babies drinking water from such contaminated sources).
- Even if domestic wastewater is treated in modern treatment plants, adding yellow water to wastewater is disadvantageous. Residual nitrogen compounds (mainly nitrate) escape with the effluents even of very efficient wastewater treatment plants and contribute to surface water eutrophication. But the main disadvantage of adding yellow water to municipal wastewater is that a great deal of energy consumption for operating activated sludge tanks is required for ammonia removal (aeration for nitrification). Removal of phosphate from municipal wastewater requires an additional biological stage and/or precipitation stages (addition of iron or aluminium salts). It is assumed that nutrient removal requires about 50 % of energy consumption in German wastewater treatment plants. It is clear that production of electric power is contributing to the greenhouse effect and leads to emission of air pollutants.
- Because of its content of nutrients, yellow water can substitute a reasonable amount of synthetic fertilizers. It is assumed that about 50 % of crops needed by human beings can be fertilized with human excreta. When yellow water is wasted, this means additional energy consumption for fertilizer production.

References

- What-wastewater, wastewater-treatment: wwdmag.com, Retrieved 15, January 2020
- What-is-black-water-why-it-is-dangerous-ignored: mwremediation.com, Retrieved 18, August 2020
- Ways-and-benefits-of-using-greywater: conserve-energy-future.com, Retrieved 02, February 2020

Chapter 2

Wastewater Management: An Introduction

Wastewater management is a combination of one or more ways through which wastewater is collected, stored, treated and finally disposed of. It is intrinsically related to the management of the water chain. Wastewater quality and quality indicators are two important features of wastewater management. This topic has been carefully written to provide a better understanding of the topic.

Wastewater is any water that requires cleaning after it is used. This includes water that has been used for laundry, bathing, dishwashing, toilets, garbage disposals, and industrial purposes. Wastewater also includes rainwater that has accumulated pollutants as it runs into oceans, lakes, and rivers. Pollutants are unwanted chemicals or materials that contaminate air, soil, and water.

The goal of wastewater management is to clean and protect water. This means that water must be clean enough so that it can be used by people for drinking and washing, and by industry for commercial purposes. It also must be clean enough to release into oceans, lakes, and rivers after it has been used.

Wastewater is usually divided into two major groups: point source wastewater and non-point source wastewater. Point source wastewater includes wastewaters that enter natural waters (such as lakes, rivers, and oceans) from defined locations. The most common point sources are sanitary sewers and storm drains. Non-point source wastewater is wastewater that is not connected to a specific source. This includes runoff (water that drains away) from agriculture and urban (city) areas, and acidic waters from mines. In many ways, point source wastewater is much easier to manage because its source and the pollutants it contains are known. Non-point source wastewater, on the other hand, is both hard to identify and treat.

Sewage Treatment

One of the largest sources of wastewater is that which comes from homes and industries. These wastewaters all flow into sanitary sewers, which direct them into sewage treatment plants. Wastewaters from homes contain human waste, food, soaps, and detergents. They also contain pathogens, which are organisms that can cause diseases. Industrial wastewaters contain toxic (poisonous) pollutants, which can endanger human health and harm other organisms. These include pesticides, polychlorinated biphenyls (PCBs,) and heavy metals like lead, mercury, and nickel. These metals are generally toxic to plant and animal life. The goal of sewage treatment is to remove all of these pollutants from the wastewater so that it can be returned to natural waters.

Sewage treatment involves three stages: primary treatment, secondary treatment, and tertiary treatment. Primary treatment physically separates solids and liquids. The wastewater passes through a grating that strains out large particles. The remaining water is left to stand in a tank, where smaller sediments (particles of sand, clay, and other materials) settle to the bottom. These sediments are called sludge. At this point, the liquid part of the wastewater still contains many pollutants and is not safe for exposure to humans or the environment.

In the second step, called secondary treatment, the liquid part of the wastewater passes through a trickling filter or an aeration tank. A trickling filter is a set of pipes with small holes in it that dribbles water over a bed of stones or corrugated plastic. Bacteria in the stones or plastic absorb pollutants from the water and break them down into substances that are not harmful. An aeration tank is a tank that contains bacteria that break down pollutants. The liquid part of the wastewater from primary treatment is pumped into the tank and mixed with the bacteria. Air is bubbled through the tank to help the bacteria grow. As bacteria accumulate, they settle to the bottom of the tank and form sludge. The sludge is removed from the bottom of the tank and buried in landfills.

After secondary treatment, the water is generally free from the majority of pathogens and heavy metals. It still contains high concentrations of nitrate and phosphate, minerals that can over-stimulate the growth of algae and plants in natural waters, which can ultimately cause them and the surrounding organisms to die. Tertiary treatment removes these nutrients from the wastewater. One method of tertiary treatment involves using biological, chemical, and physical processes to remove these nutrients. Another method is to pass the water through a wetland or lagoon (shallow body of water cut off from a larger body).

Storm Sewers

In most cities in the United States, the sewers that carry storm waters are routed through sewage treatment plants. Much of the runoff from storms contains fertilizers, oils, and other chemicals that should be removed from the water before it enters lakes, rivers, and oceans. When there are very heavy rainfalls, however, the sewage treatment plants can become overwhelmed by the volume of water entering the facility. At these times, sewage and wastewater from storms may be dumped directly into natural water bodies. Many cities have programs underway to separate the storm sewers from sanitary sewers, but these projects are very costly and time consuming.

Agricultural Runoff

Agricultural runoff occurs when rain falls to the ground and then runs through agricultural fields or livestock-raising farms. The rainwater can accumulate fertilizer, oils, and animal wastes before it runs into rivers, lakes, and oceans. These materials pollute natural waters and can cause fish to die, contaminate drinking water, and speed up the rate of sedimentation (particles settling to the bottom of a waterway) in lakes and streams. In the summer of 1995, runoff from hog farms in North Carolina caused the rapid growth of the algae Pfisteria. This algae released toxins that affected the nervous system of fish as well as humans in the area.

In an attempt to manage agricultural runoff, the Office of Wastewater Management (OWM) of the U.S. Environmental Protection Agency has designated farms as Animal Feeding Operations (AFOS). As of 1998, nearly half a million AFOS had been identified. By designating AFOS, the

OWM can regulate the disposal of animal waste products. This moves a large portion of agricultural runoff from the non-point source category to the point source category, and allows for better management of agricultural pollutants.

Acid Mine Drainage

In places where coal is mined, the mineral pyrite is a waste material. A series of complex reactions between pyrite, oxygen, and water result in acid mine drainage. Acid mine drainage is wastewater that is extremely acidic and contains high concentrations of heavy metals. Acid mine drainage is one of the major sources of stream pollution in the Appalachian mountain region. Acid mine drainage has severely damaged more than half the streams in Pennsylvania and West Virginia. There are at least 200,000 abandoned mines throughout the United States that produce acid mine drainage.

Acid mine drainage can be treated using chemical treatments that decrease the acidity of the water, and allow the heavy metals to precipitate (separate from the water). This type of treatment is often very expensive. Another way to treat acid mine drainage is by passing it through a lagoon or wetlands, which removes heavy metals and decreases the acidity of the water. Acid mine drainage is also treated by passing it through a channel of limestone (a rock that is very alkaline), which also neutralizes the acidity of the water.

Urban Runoff

When rain falls on natural lands such as forests and meadows, some of it soaks into the soil and then slowly makes its way to rivers, lakes, and oceans. In cities, much of the land is paved with cement and asphalt, and water is unable to sink into the ground. Instead, it quickly moves to storm drains and then into natural waterways. This great volume of water causes much erosion (wearing away of the land) and sedimentation. In addition, as the rainwater runs over paved surfaces, it gathers oil and grease from cars, fertilizers and pesticides from gardening, pathogens form animal wastes, road salts, and heavy metals. These are dumped directly into natural waters with urban wastewater. Runoff from urban areas is the largest source of pollution in estuaries (the wide part of a river where it nears the sea) and the third largest source of pollution in lakes.

Controlling urban runoff is extremely difficult because its sources are hard to identify. Some ideas to minimize runoff include adding vegetation and drainage areas to new construction sites. Some cities have instituted sewer-stenciling programs that remind people that rainwater flows directly into natural waters. Gas stations have also been targeted as businesses that can help control car oils and grease. Schools have also developed programs to teach students about urban runoff and non-point source wastewater.

Wastewater Quality

Good water quality is essential to human health, social and economic development, and the ecosystem. However, as populations grow and natural environments become degraded, ensuring there are sufficient and safe water supplies for everyone is becoming increasingly challenging.

A major part of the solution is to produce less pollution and improve the way we manage wastewater.

A more circular and therefore more sustainable economy requires us to value wastewater for its potential, rather than discard or ignore it. More than just an alternative source of water, safe wastewater management could help protect our ecosystems and give us energy, nutrients and other recoverable materials.

Challenges and Opportunities

Water must be carefully managed during every part of the water cycle: from fresh water abstraction, pre-treatment, distribution, use, collection and post-treatment, to the use of treated wastewater and its ultimate return to the environment, ready to be abstracted to start the cycle again. Due to population growth, accelerated urbanization and economic development, the quantity of wastewater generated and its overall pollution load are increasing globally.

The availability of safe and sufficient water supplies is inextricably linked to how wastewater is managed. Increased amounts of untreated sewage, combined with agricultural runoff and industrial discharge, have degraded water quality and contaminated water resources around the world. Globally, 80% of wastewater flows back into the ecosystem without being treated or reused, contributing to a situation where around 1.8 billion people use a source of drinking water contaminated with faeces, putting them at risk of contracting cholera, dysentery, typhoid and polio. Far from being something to discard or ignore, wastewater will play a major role in meeting the growing water demand in rapidly expanding cities, enhancing energy production and industrial development, and supporting sustainable agriculture.

Figure: Contaminated water in Karial slum, one of the urban.

Wastewater and Cities

Mainly in low-income areas of cities and towns within developing countries, a large proportion of wastewater is discharged directly into the closest surface water drain or informal drainage channel, sometime without or with very little treatment. In addition to household effluent and human waste, urban-based hospitals and industries such as small-scale mining and motor garages, often dump highly toxic chemicals and medical waste into the wastewater system.

Even in cities where wastewater is collected and treated, the efficiency of treatment may vary according to the system used. However, water used by a municipal authority for irrigating green spaces or cleaning streets does not need to be treated to a potable standard. Treating wastewater to a water quality standard appropriate to its intended use increases the potential for cost recovery.

The growth of urban demand for water will require new approaches to wastewater collection and management. Indeed, reused wastewater may help address other challenges including food production and industrial development.

Wastewater and Industry

Societal and environmental pressures over recent years have led to a growing movement for industry to reduce its wastewater and to treat it before discharge. Wastewater is now seen as a potential resource and its use, or recycling after suitable treatment, can provide economic and financial benefits. Societal and environmental pressures over recent years have led to a growing movement for industry to reduce its wastewater and to treat it before discharge. Wastewater is now seen as a potential resource and its use, or recycling after suitable treatment, can provide economic and financial benefits.

Wastewater can be used within the business itself or between several businesses through 'industrial symbiosis'. Industrial water consumption is responsible for 22% of global water use. In 2009 in Europe and North America, water consumption by industries was 50% as compared to 4-12% in developing countries. It is expected that in rapidly industrialising countries, this proportion could increase by a factor of five in the next 10-20 years. Therefore, there is a strong incentive to use wastewater in-house and locally, based on cost savings alone.

Figure: Wastewater and industry.

Businesses can directly use some wastewater, providing it is fit for purpose. For instance, using process water for cooling or heating, or rainwater from roof collection or concrete aprons for toilet flushing, irrigation or vehicle washing.

Wastewater in Agriculture

Partly to help maximise yields to meet demand, usage of chemical fertilizers and pesticides has increased in recent years both in industrial and small farming, making agriculture a potential source of environmental pollution.

Pollution of groundwater and surface water by agricultural use of untreated or inadequately treated wastewater is a major issue in many developing countries where such irrigation is practised. Improved wastewater management can improve the health of workers, especially in agriculture, by reducing the risk of pathogen exposure.

Farmers are increasingly looking into non-conventional water resources, mainly wastewater, whether due to its high nutrient content or lack of conventional water resources. If applied safely, wastewater is a valuable source of both water and nutrients, contributing to water and food security and livelihood improvements.

Figure: Farmers are increasingly looking into non-conventional water resources, mainly wastewater, whether due to its high nutrient content or lack of conventional water resources.

Wastewater Quality Indicators

Water is an essential component of life and we probably don't think about where our water comes from until it is not there. The water we use is constantly recycled — by nature using the water cycle and by humans using our treatment plants — it is a finite resource, we cannot make water.

Water is known as the universal solvent because of its ability to dissolve a vast array of substances. Whilst this makes it ideal as a solvent for cleaning our clothes — it does mean that wastewater is good at dissolving compounds harmful to the environment. Strict regulations govern the discharge of wastewater to our rivers and streams.

Laboratory analysis is used to assess the quality of the wastewater and to determine if it is safe to be discharged back into the environment — but there are natural indicators as to the cleanliness of a water supply.

Streams, rivers and lakes produce and consume oxygen naturally. At a constant temperature a healthy water supply would be in oxygen equilibrium — gaining oxygen from the atmosphere and from aquatic plants due to photosynthesis, and losing oxygen due to respiration by aquatic animals, chemical reactions in the water and decomposition.

Wastewater often contains organic materials that can be decomposed by microorganisms, using oxygen as this happens. If the wastewater contains too much organic material and is discharged into a stream it can cause a drop in the oxygen content of the stream as the microorganisms decompose the organic material.

The amount of oxygen consumed by the microorganisms in breaking down the waste is known as the BOD. So by monitoring the BOD of the wastewater, we can make sure that too much organic material is not discharged.

This is similar to BOD in that it measures how much oxygen will be removed from the clean water by the discharge of wastewater, but this considers the reactions of oxidizable chemicals (nitrates and sulfates) rather than organic matter.

Physical Characteristics

Two of the physical criteria that have to be met before wastewater can be discharged back into the environment are:

- Temperature - Discharged water must be at a temperature close to the temperature of the receiving water as many organisms can only survive within a narrow temperature range.
- Dissolved Solids - Refers to material that can be dissolved or suspended in the wastewater. Dissolved solids can indirectly and directly affect aquatic animals by reducing light levels and clogging gills in fish.

Biological Oxygen Demand

Biochemical oxygen demand or biological oxygen demand represents the amount of oxygen consumed by bacteria and other microorganisms while they decompose organic matter under aerobic (oxygen is present) conditions at a specified temperature.

When you look at water in a lake the one thing you don't see is oxygen. In a way, we think that water is the opposite of air, but the common lake or stream does contain small amounts of oxygen, in the form of dissolved oxygen. Although the amount of dissolved oxygen is small, up to about ten molecules of oxygen per million of water, it is a crucial component of natural water bodies; the presence of a sufficient concentration of dissolved oxygen is critical to maintaining the aquatic life and aesthetic quality of streams and lakes.

The presence of a sufficient concentration of dissolved oxygen is critical to maintaining the aquatic life and aesthetic quality of streams and lakes. Determining how organic matter affects the concentration of dissolved oxygen (DO) in a stream or lake is integral to water- quality management. The decay of organic matter in water is measured as biochemical or chemical oxygen demand. Oxygen demand is a measure of the amount of oxidizable substances in a water sample that can lower DO concentrations.

Certain environmental stresses (hot summer temperatures) and other human-induced factors (introduction of excess fertilizers to a water body) can lessen the amount of dissolved oxygen in a water body, resulting in stresses on the local aquatic life. One water analysis that is utilized in order to better understand the effect of bacteria and other microorganisms on the amount of oxygen they consume as they decompose organic matter under aerobic (oxygen is present) is the measure of biochemical oxygen demand (BOD).

Determining how organic matter affects the concentration of dissolved oxygen in a stream or lake is integral to water-quality management. BOD is a measure of the amount of oxygen required to remove waste organic matter from water in the process of decomposition by aerobic bacteria (those bacteria that live only in an environment containing oxygen). The waste organic matter is stabilized or made unobjectionable through its decomposition by living bacterial organisms which need oxygen to do their work. BOD is used, often in wastewater-treatment plants, as an index of the degree of organic pollution in water.

Chemical Oxygen Demand

Oxygen demand is an important parameter for determining the amount of organic pollution in water. Testing oxygen demand has its widest application in measuring waste loadings of treatment plants and in evaluating treatment efficiency. Estimating BOD is an important part of wastewater treatment process control. High influent BOD requires extensive treatment to provide the oxygen necessary to break down the water's organic contents.

Left untreated or partially treated, discharged water contains effluent organics that compete with downstream organisms for oxygen. This oxygen demand can kill or inhibit life downstream of the discharge area.

The U.S. Environmental Protection Agency (USEPA) requires wastewater treatment plants to bring BOD within limits before discharging treated wastewater, and accurate test results must be recorded for regulatory reporting. Failure to do so is a violation of federal law that can lead to extensive fines. Thus, measuring BOD in treated water is an important part of the monitoring process. In addition, plants that treat wastewater from commercial operations measure the oxygen demand of waste as it comes to the facility to determine how much the commercial customer must pay in fees to have its waste treated.

By monitoring wastewater entering the treatment facility, technicians are able to respond to changes in oxygen demand and adjust the treatment process accordingly. The challenge of this process lies in the fact that the BOD of incoming wastewater can vary substantially over days or hours, and accurate BOD measurement takes five days to complete.

BOD testing uses microorganisms that consume oxygen while feeding on organic compounds in a wastewater sample over a five-day period. While this test is a good model of the aerobic waste treatment process, in some cases the microorganisms can become poisoned by toxic substances in the untreated wastewater. The five-day test also does not provide the real-time information necessary to make process control decisions.

For this reason, many wastewater treatment facilities use a faster Chemical Oxygen Demand (COD) test to estimate BOD levels. The EPA-approved COD test is useful for performing rapid, frequent monitoring of treatment plant efficiency, and results allow quick response to changing conditions in the waste stream. The evaluation is complete in just two hours, and the reproducible results correlate with BOD and often can be used for NPDES reporting. Toxic materials in the sample do not affect the oxidant, so the test provides a good indicator of organic pollution in industrial wastewater containing heavy metals and cyanides.

COD testing assesses all chemically oxidizable substances and can be directly related to the true oxygen demand imposed by the effluent if released into the environment. Because each organic compound differs in the amount of oxygen necessary for complete oxidization, the COD test reflects the effect of an effluent on the receiving stream more directly than measurement of carbon content.

Total Organic Carbon testing can be performed as a complementary analysis method to COD. The COD/TOC ratio can be an important tool in the evaluation of wastewater treatment. While the TOC test directly assesses the carbon atoms present in organic compounds, operators of processes dependent upon on biochemical oxidation (e.g. wastewater treatment) find COD provides a measure of the oxygen-demanding substances. The ratio of COD value to TOC value at a given point in the process may provide information on the type of organic wastewater constituents present.

A high COD/TOC ratio may indicate organic compounds that are easily oxidized (alcohols, for example). A shift in the COD/TOC ratio in the influent means a change in the type of organic compounds entering the system, which can impact the effectiveness of the process. In the wastewater treatment, the amount of oxygen required (reflected by the COD value) may change while the carbon concentration (reflected by the TOC value) does not.

By-products of the traditional dichromate COD test, including mercury, chromium and silver, can be costly and troublesome to dispose of. A patented COD test available from Hach uses manganese

III oxidant instead of the heavy metal reagents used in the traditional test. This methodology allows the fast, accurate estimation of BOD, and the end products do not contain toxic metals. Mercury-free dichromate COD reagents are also available and help minimize disposal costs.

Users should be aware that, with mercury-free alternatives, some organic compounds are not oxidized completely, and chloride ions cause interference with the oxidation process, necessitating their removal for accurate results. The Hach Manganese III method includes a chloride pretreatment method for samples with interfering chloride levels.

COD testing, via the dichromate or mercury-free method, is an efficient alternative or complementary method to the BOD test for wastewater monitoring. Its relatively quick turnaround time, simplicity and correlation to BOD make it a valuable tool for those analyzing wastewater for pretreatment planning or process monitoring.

Wastewater Treatment Plant

A wastewater treatment plant is a facility in which a combination of various processes (e.g., physical, chemical and biological) are used to treat industrial wastewater and remove pollutants.

Effluent Treatment Plants (ETP)

Effluent Treatment Plants or (ETPs) are used by leading companies in the pharmaceutical and chemical industry to purify water and remove any toxic and non toxic materials or chemicals from it. These plants are used by all companies for environment protection.

An ETP is a plant where the treatment of industrial effluents and waste waters is done. The ETP plants are used widely in industrial sector, for example, pharmaceutical industry, to remove the effluents from the bulk drugs.

During the manufacturing process of drugs, varied effluents and contaminants are produced. The effluent treatment plants are used in the removal of high amount of organics, debris, dirt, grit, pollution, toxic, non toxic materials, polymers etc. from drugs and other medicated stuff. The ETP plants use evaporation and drying methods, and other auxiliary techniques such as centrifuging, filtration, incineration for chemical processing and effluent treatment.

The treatment of effluents is essential to prevent pollution of the receiving water. The effluent water treatment plants are installed to reduce the possibility of pollution; biodegradable organics if left unsolved, the levels of contamination in the process of purification could damage bacterial treatment beds and lead to pollution of controlled waters. The ETPs can be established in the industrial sectors like Pharmaceuticals, Chemicals and Leather industry and tanneries.

Sewage Treatment Plants (STP)

Sewage treatment, or domestic wastewater treatment, is the process of removing contaminants from wastewater and household sewage, both runoff (effluents) and domestic. It includes physical, chemical, and biological processes to remove physical, chemical and biological contaminants.

Its objective is to produce a waste stream (or treated effluent) and a solid waste or sludge suitable for discharge or reuse back into the environment. This material is often inadvertently contaminated with many toxic organic and inorganic compounds.

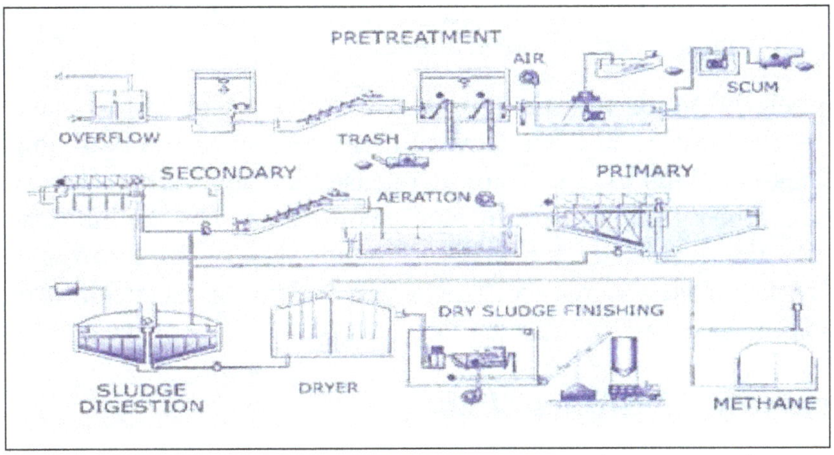

Figure: Process Flow Diagram for a typical large-scale treatment plant.

Pre-treatment removes materials that can be easily collected from the raw wastewater before they damage or clog the pumps and skimmers of primary treatment clarifiers, for example, trash, tree limbs, leaves, etc.,

The influent sewage water is strained to remove all large objects carried in the sewage stream. This is most commonly done with an automated mechanically raked bar screen in modern plants serving large populations, whilst in smaller or less modern plants a manually cleaned screen may be used this is called as screening.

The raking action of a mechanical bar screen is typically paced according to the accumulation on the bar screens and/or flow rate. The solids are collected and later disposed in a landfill or incinerated. Pre-treatment may include Grit removal in which, a sand or grit channel or chamber where the velocity of the incoming wastewater is carefully controlled to allow sand, grit and stones to settle.

Common and Combined Effluent Treatment Plants (CETP)

Many of the Small Scale Industries (SSI) are unable to put up the treatment systems individually, the concept of CETP's (Common Effluent Treatment Plants) is envisaged to benefit such industries in treating its effluent before disposal whether it is in stream, land, sewerage system or in rivers and seas. CETP's are set up in the industrial estates where there are clusters of small scale industrial units and where many polluting industries are located.

The Ministry of Environment & Forest has launched the centrally sponsored scheme, namely, Common Effluent Treatment Plant (CETP) in order to make a cooperative movement of pollution control especially to treat the effluent, emanating from the clusters of compatible Small-Scale Industries. The major objective of the CETP is therefore, to reduce the treatment cost to be borne by an individual member unit to a maximum while protecting the water environment to a maximum.

The proposal for setting up of CETP's by such industries is to be submitted by the CETP Association to the respective State Pollution Control Board, which after examining the proposal and obtaining commitment from the concerned State Government regarding its contribution will give their recommendation to the Ministry of Environment and Forests for consideration , the Ministry examines the proposal and takes the decision through a Screening Committee constituted in this regard for providing support from the Central Government.

The Ministry releases the funds for the approved projects which is the matching grant to the amount released by the concerned State Government, subject to the bank guarantee to be taken from the CETP associations for the amount released by the Central Government, The CETP Company should meet the remaining cost by equity contribution by the industries and loans from financial institutions. Funds released for the CETP's should be utilized for the CETP only and not for payment for any debts/bank loans, etc. The design and technical specification of CETP can be referred in any book on wastewater management.

Design of Wastewater Treatment Plant

A wastewater treatment plant is to be designed for a wastewater with the following information. N is the average of last digits of two team members. Design questions are inserted in between the information for easiness:

- Some bigger size impurities such as bottles and hair are present.
- Silt particles of 0.0170*(1+N*0.1) cm diameter and 3*(1-N*0.1) g/cm^3 are present. Design a grit chamber. What are the advantages of an aerated grit chamber?
- Obtain inflow rate and BOD5 results in page 529 for question 6-7 by multiplying the flow rate by (1+N*0.1) and BOD5 by (1+N*0.05).
- Design an equalization basin.
- 60% removal efficiency of suspended solids should be achieved in primary sedimentation tank. Use Figure P6.12 on page 530 of the book for settlement characteristics. Design the primary sedimentation tank: calculate the necessary overflow rate and thus estimate the surface area required then calculate the dimensions. Design the overflow weir using assumed (from the book) weir loading rates.
- Microbes growing in the aeration tank are assumed to have the characteristics. K_s, 40*(1+0.1*N) mg BOD_5/L; k_d = 0.1*(1+0.1*N) d^{-1}; μm = 2*(1+0.1*N) d^{-1}; Y=0.4*(1+0.05*N) mg VSS/mg BOD_5.
- 35+20*(1+0.02*N)% of the BOD5 in the wastewater treatment plant effluent can be assumed to be in ss form.
- In the activated sludge about 40% of the mixed liquor suspended solids is non volatile.
- Calculate the mean cell residence time.
- If MLVSS in the aeration tank is expected to be 2000*(1+N*0.05) mg/L, calculate the volume of aeration tank, hydraulic retention time, F/M ratio (check if is feasible), return sludge flow rate, sludge production and oxygen requirement.

- From your own reading, design the secondary settling tank.
- Form your own reading of the book, what are possible means of treating the waste sludge?
- What are the advantages and pitfalls of biological phosphorus removal process?

Designing

Grit Chamber

N=5

Silt particles diameter = 0.017*(1+N*0.1) = 0.0255 cm

ρs = 3*(1-N*0.1) = 1500 kg/m³

Note that the product of the specific gravity of the particle (1.5) and the density of water is the density of the particles (ρ_s).

Before we can calculate the terminal settling velocity of the particle at wastewater temperature of 22 °C, we find the water density to be 997.774 kg/m³. We will use 1000 kg/m³ as a sufficiently close approximation. We find the viscosity to be 0.995 mPa. Silt particles diameter is 0.000255 m:

$$V_s = \frac{g(\rho s - \rho)d2}{18\mu}$$

$$V_s = \frac{9.8(1500-1000)0.000255}{18(0.000995)} = 0.0177\,\mathrm{m/s} = 17.7\,\mathrm{mm/s}$$

The Reynolds number for this settling velocity (0.0177 m/s) and particle size is:

Re = (0.0177)*(0.000255)/(0.000995/1000) = 4.54

From the iterative solution Vo = 0.028 m/s. with a flow of 0.15 m³/s and a horizontal velocity of 0.25 m/s the cross sectional area of flow may be estimated to be:

Ac = (0.15 m³/s)/(0.25m/s)= 0.60 m²

The depth of flow is then estimated by diving the cross-sectional area by the width of the channel:

h = 0.60 m²/0.56m =1.07m

If the grit particle in question enters the grit chamber at the liquid surface, it will take h/Vs seconds to reach the bottom:

t = 1.07m/0.028m/s = 38.2 s

Since the chamber is 13.5 m in length and the horizontal velocity is 0.25 m/s the liquid remains in the chamber:

t = 13.5m/0.025m/s= 54s

Thus, the particle will be captured in the grit chamber:

Overflow velocity = 0.15/(13.5*0.56)=19.8 mm/s

Vs/Vo = 17.7 / 19.8 = 0.893 < 1 hence some sand particles with this diameter and density would settle down.

The assumption is that the horizontal grit chamber that is 13.5 m in length if the average grit chamber flow is 0.15 m^3/s, the width of the chamber is 0.56 m and the horizontal velocity is 0.25 m/s.

Advantages

Girt removal from wastewater happens through Aerated Grit Chamber removes grit. Grit chamber has different advantages:

- Possible septic conditions of the plant influent may be alleviated through pre-preparation in the grit chamber.
- Consistent removal efficiency over a wide flow range.
- Performance of downstream units may be improved by using pre-aeration to reduce septic conditions in incoming wastewater.
- Aerated grit chambers are versatile, allowing for chemical addition, mixing, pre-aeration and flocculation.
- Maintenance is significantly reduced.
- The mechanical design is very simple.
- Tin this process there is no moving parts under the water surface.
- Blower air can be used in this method to air-lift pumping.
- A relatively low decayable organic content may be removed with a well-controlled rate of aeration.

Equalization Basin

Flow equalization is not a treatment process in itself, but a technique that can be used to improve the effectiveness of both secondary and advanced wastewater treatment processes.

The first step then is to calculate the average flow. In this case it is 0.1404 m^3/s. Next the flows are arranged in order beginning with the time and flow that first exceeds the average. In this case it is at 0900h with a flow of 0.1965 m^3/s.

Explanation of the calculations for each column follows:

Volume inflow = (0.1965 m^3/s)(1h)(3600 s/h) = 707.4 m^3

Same step for all value:

Volume outflow = (0.1404 m^3/s) (1h) (3600s/h) = 505.44 m^3

Difference between the inflow volume and outflow volume:

dS = Vin – Vout

The cumulative sum of the difference between the inflow volume and outflow volume:

Storage = $\sum$dS

The required volume for the equalization basin is the maximum cumulative storage. With the requirement for 25 percent excess, the volume would then be:

Storage volume = 2219.4m^3*1.25 = 27774.25 m^3

The average concentration is determine as:

$$Savg = \frac{(Vi)(So) + (Vs)(Sprev)}{Vi + Vs}$$

Where, Vi =volume of inflow during time interval Δt, m^3.

So = average BOD5 concentration during time interval Δt, g/m^3.

Vs = volume of wastewater in the basin at the end of the previous time interval m^3.

Sprev = concentration of BOD5 in the basin at the end of the previous time interval g/m^3.

Primary Sedimentation

Evaluate the following primary tank design with respect to detention time, overflow rate, and weir loading.

Design data:

Flow = 0.1965 m^3/s.

Influent ss = 286 mg/l.

Sludge concentration = 6 %.

Efficiency = 60%.

Length = 40 m (effective).

Width = 10 m liquid depth = 2 m.

Weir length = 75m.

The detention time is simply the volume of the tank divided by the flow:

θ = V/Q = (40*10*2)/ 0.1965 = 4071.24 s = 1.13 h

This is a reasonable detention time.

The overflow rate is the flow divided by the surface area.

Vo = 0.1965 / (40*10) = 0.000491 m/s x 86400 s/d = 42.4 m/d

This is an acceptable overflow rate.

The weir loading is calculated in the same fashion:

WL = 0.1965/75 = 0.00262 m^3/s. m x 86400 s/d = 226.368 m^3/d. m

This is an acceptable weir loading:

Ks = half velocity constant, mg/l = 60 mg BOD5 /L.

Kd = decay rate of microorganisms, d^{-1} = 0.15 d^{-1}.

μm = maximum growth rate constant, d^{-1} = 3 d^{-1}.

Y = decimal fraction of food mass converted to biomass, mg VSS/mg BOD5 = 0.5 mg VSS/mg BOD5.

BOD5 = 84 mg/L

Flow = 0.150 m^3/s

57% of the BOD5 in the wastewater treatment plant effluent can be assumed to be in SS form = (84*57)/ 100 = 47.88 mg/l = 48 mg/L

In the activated sludge about 40% of the mixed liquor suspended solids is non-volatile:

S = BOD5 allowed – BOD5 in suspended solids

S = 48 – (0.4)(48) = 29 mg/L

The mean cell-residence time can be estimated with equation: $S=\frac{Ks(1+KdOc)}{Oc(\mu m-Kd)-1}$ and the assumed values for the growth constants: $29=\frac{60(1+0.15Oc)}{Oc(3-0.15)-1}$

Solving for θc:

(29)(2.85θc – 1) = 60 + 9θc

82.65 θc – 29 = 60 + 9 θc

$$\theta c=\frac{89}{73.65}=1.2d$$

If we assume a value of 2000*(1+N*0.05) mg/L for the MLVSS, we can solve equation $X=\frac{\theta c(Y)(So-S)}{\theta(1+Kd\theta c)}$ for the hydraulic detention time.

$$2500 = \frac{1.2(0.5)(84-29)}{O(1+0.15*1.2)}$$

$$\theta = \frac{33}{2950} = 0.01d \text{ or } 0.26h$$

The volume of the aeration tank is then estimated using equation: $\frac{V}{Q} = \theta$

$$0.26 = \frac{V}{0.150 \times 3600}$$

$$V = 140.4m^3 \text{ or } 140m^3$$

References

- Quality-and-wastewater-2, water-facts: unwater.org, Retrieved 05, July 2020
- An-introduction-to-wastewater-quality-indicators, water-wastewater-33269: pollutionsolutions-online.com, Retrieved 29, March 2020
- Nuts-and-bolts-chemical-oxygen-demand: wqpmag.com, Retrieved 04, June 2020

Chapter 3

Wastewater Pollution

Wastewater pollution refers to the contamination of water bodies, land and ground water by wastewater discharge. Wastewater contains many harmful chemicals and substances that render water unfit for consumption and other purposes, and even affects the soil. This chapter sheds light on wastewater pollution and its effects on the different systems.

Pollution of industrial waste water is caused by the discharge of harmful chemicals and substances into water, making it unsuitable for consumption and for other purposes. Although 60 to 70% of the land is covered with water, so water bodies such as lakes, ponds, rivers, reservoirs and watercourses provide us with fresh water, keep it clean, it is a matter of survival, not only for people, but also for all other life forms. Since the industrial revolution, we have achieved many things; our manufacturing process to become more efficient and productive, science has become more sophisticated and our lives have changed. But perhaps nothing comes without a price. All the progress and development seen in past centuries have also brought with them a wide range of problems; water pollution is one of them. Fouling refers to pollution and hazardous waste environmental materials, which lead to a significant change in the quality of the surrounding atmosphere. Pollution of the environment can be classified as air pollution, water pollution and noise. Water pollution means water pollution, which is unfit for consumption and for other purposes. There are essentially five main sources of water pollution from domestic wastewater, agricultural run-off and industrial wastewater, sewage from septic tanks and rainwater. There is a lack of stringent pollution control policies in many countries around the world, particularly in developing and underdeveloped countries. In many countries, policies exist, but the apathy of law enforcement authorities has allowed industries to take these laws for granted and easily circumvent them.

Some industries still rely on old and obsolete technologies that produce more pollutants than modern technologies. Industries essentially try to avoid the high cost of modern or sophisticated technologies by using outdated technologies, although they are known to be less efficient than modern technologies. In many countries, industrial water is not treated adequately before it is discharged into rivers or lakes. This is especially true for small industries that do not have enough capital to invest in pollution control equipment. Unplanned industrial growth contributes to water pollution. Although industrial growth stimulates a country's economy, it degrades the environment, especially when it is unforeseen. Growth can also contribute to limiting waste disposal sites and total disregard for pollution control laws are negative consequences of unplanned industrial growth. The extraction of minerals by mining and drilling causes pollution of industrial water. Extraction and drilling operations render the land unusable for agricultural activities and any accidental leakage can escape into the surrounding water and enter the ocean. Oil spills can pollute land and sea. Waste generated during mining operations can increase the salt and mineral content of water and change its pH. Mine mining pollutes both surface water and groundwater.

These wastes generated in various industrial processes can basically achieve these changes, when poured into water bodies. Industrial water pollution can have far-reaching consequences for the ecosystem. Water is used in various industrial processes, coming into contact with toxic substances, heavy metals, organic sludge and even radioactive sludge. So, when such polluted water thrown into the ocean or other bodies of water without any treatment, they become unsuitable for any human and agricultural use. Radioactive sludge deposited on the bottom of bodies of water can remain highly radioactive for many decades and pose a serious health risk to people living nearby.

Nuclear reactors are also a major source of pollution of the heat along with power. Thermal pollution refers to an increase in ambient temperature. This may have an adverse effect on aquatic or marine life, as some organisms are extremely sensitive to small changes in temperature. If the nutrient content of water is undergoing changes can disrupt the delicate balance of the ecosystem. For example, if the content of nutrients increases the water, which is known as eutrophication, may promote water bloom. Algal bloom can deplete oxygen in the water. When algae produce oxygen during the day with photosynthesis, at night, which use oxygen dissolved in water. Water bloom is often followed by death of large amounts of algae, which are decomposed by bacteria using oxygen. This means that the whole process of algae blooms and subsequent bacterial decomposition of algae deplete the dissolved oxygen in the water. Sometimes this process can reduce the content of oxygen in the water to such an extent that it can no longer support aquatic life. Such hypoxic areas of the ocean where marine life cannot survive are known as dead zones. Common industrial pollutants responsible for causing water pollution are sulphur; asbestos, noxious solvents, polychlorinated biphenyls, lead, mercury, nitrates, phosphates, acids, alkali, dyes, pesticides, benzene, chlorobenzene, carbon tetrachloride, toluene and volatile organic chemicals such as sulphur are harmful to marine life, while asbestos is known to be a potential carcinogen. Drinking water contaminated with asbestos may increase the risk of benign colon polyps.

Nitrates and phosphates are essentially found in fertilizers and may contribute to the eutrophication process that can cause dead zones. On the other hand, the presence of high levels of carbon tetrachloride in drinking water may cause problems with the liver. Other industrial pollutants benzene is suspected to be associated with diseases such as anemia and decreased platelet counts and increased risk of cancer. Chemical chlorobenzene found in insecticides and colorants may adversely affect on the liver, kidney and central nervous system. Toluene is substantially pollutant generated by the oil and petroleum industry, which can also affect the liver, kidney and central nervous system. Volatile organic compounds are essentially solvents used in a wide variety of industrial and household products. If not properly disposed of, these chemicals contaminate groundwater. They can cause a variety of health problems such as headaches, nausea, liver and kidney damage and memory. When industrial water pollution, it is difficult to contain, it is not impossible. Greater awareness should be created among the common masses about how water gets polluted, it can be prevented its effects on human health and marine life and how. It is not possible to reduce water pollution without public cooperation and cooperation of industrial units. Strict pollution control laws and regulations and their effective implementation have an important role in controlling any kind of pollution. Development of cost-effective pollution control equipment and incentives from the government for installing these devices can encourage industry to take control pollution seriously.

Some industrial wastes are common waste, as well as domestic sewage. Such waste can be easily treated municipal facilities. But some wastes, such as heavy metals, volatile organic compounds and oils and fats need special treatment. Industries can install separate pre-treatment of such hazardous waste. Partially treated waste water may be sent to the local system for further purification. Extensive industries generate large amounts of waste water. These sectors should revise their manufacturing processes to reduce the amount of pollutants and operate their own facilities for treatment on the spot. Industrial wastewater treatment can be performed in three stages-primary treatment, which includes mechanical processes, biological processes for secondary treatment and tertiary treatment which can be performed using biological, physical and chemical processes. In the primary treatment, pollutants are separated from the water by screening, grinding, flocculation and sedimentation processes. In the secondary treatment, biological methods are used for wastewater treatment. Finally, the waste water is recycled to the tertiary treatment using biological, physical and chemical processes. Thermal pollution on the other hand, can be controlled by creating cooling ponds, or by using a cooling tower. The cooperation of citizens and industrial units, proper enforcement of legislation and the availability of effective and cheaper pollution control techniques are essential prerequisites for the success of such legislation.

Effects of Wastewater Pollution

Effects on Aquatic Ecosystems

Wastewater and sewage enter aquatic systems from sources ranging from surface runoff and septic systems to wastewater treatment facilities and storm drain outfalls. Every year about 3.5 million Americans get sick from recreational activities like swimming and boating because the water is contaminated. Many don't connect their illness with the water they touched. The impact of water pollution on aquatic ecosystems extends far beyond human illness, however.

Wastewater treatment has three phases. The first phase or primary treatment places the wastewater in holding ponds. Solid waste settles at the bottom, and low-density materials like fats and oils float to the top. These materials can then be removed. The second phase or secondary treatment removes dissolved and suspended biological material. Most secondary treatment systems use aerobic bacteria to consume organic material in the wastewater. Tertiary or third-phase treatment further cleans wastewater that will eventually be released into sensitive environments. Tertiary treatment can be accomplished by several methods, depending on the remaining contaminants. Sand filtration removes particulate matter. Phosphates may be removed using bacteria called polyphosphate accumulate organisms. Nitrifying bacteria can be used to remove nitrogen. A method called lagooning stores the water in a lagoon where plants, bacteria, algae and zooplankton consume the remaining contaminants through natural processes.

The solid waste called sludge removed during primary treatment receives secondary treatment as well. The sludge may be treated with bacteria. Sometimes the bacteria generate enough methane to be used as fuel. Or, the sludge can be incinerated. Another method to treat the sludge starts by condensing the sludge, heating it to disinfect it and then finally using the treated sludge as fertilizer.

Despite the Clean Water Act of 1972 requiring secondary treatment of wastewater, some U.S.

municipalities filed and received exemptions. Around the world, an estimated 2.5 billion people lack improved sanitation facilities. Increasing populations, aging infrastructure and natural disasters also impact the effectiveness of wastewater treatment systems.

Domestic wastewater contains pollutants ranging from biologic hazards and microplastic particles to soaps and fats. Agricultural wastewater contains biologic hazards, salts, pesticides and fertilizers. Urban wastewater includes domestic and industrial wastewater but also contains runoff from storm drains. Storm drains carry pollutants from yards and parks (dirt, pet waste, pesticides, herbicides and fertilizers) as well as from streets and parking lots (oil, gasoline, dirt and trash). Industrial wastewater contains a wide range of chemicals that include petrochemicals and other chemicals, acids, radioactive materials and salts. Recent findings show that a variety of drugs also contaminate wastewaters.

Biologic Hazards in Aquatic Environments

Biologic hazards found in wastewater include bacteria, fungi, parasites and viruses. Bacteria and bacterial diseases range from E. coli, typhoid fever, salmonella, cholera and shigellosis. Fungi include aspergillus. Parasites include cryptosporidium, giardia and roundworms. Viruses like hepatitis A can also be found in wastewater. Health problems caused by sewage pollution impact an estimated 3.5 million Americans each year. An estimated 50 percent of the wastewater entering the Mediterranean is untreated sewage. Biologic waste from farms, houses, parks and beaches cause health problems that impact more than humans.

Bacteria and other organisms in freshwater use oxygen to metabolize the sewage they accompany. While breaking down the sewage, these micro-organisms can cause hypoxic (oxygen-depleted) dead zones. These dead zones lack oxygen that fish and other native organisms need to survive. Shellfish infected with sewage-related bacteria sicken people around the world. In marine environments, human gut bacteria can infect coral and cause coral bleaching disease. When coral lose their natural bacteria and algae, they die, resulting in zones where the coral ecosystem, from bacteria to fish populations, dies.

Drugs ranging from hormones (which affect reproductive development in fish and amphibians) to legal and illegal amphetamines to antidepressants have entered aquatic ecosystems. Some of the drugs pass into the sewage system in the urine and feces of users while some of the drugs have been flushed down the drain. One controlled study of the effects of amphetamines on aquatic organisms showed accelerated insect reproduction, decreased algae populations and changes in diatom and microbe diversity.

Nutrient Hazards in Aquatic Environments

Nutrient-rich materials from fertilizers, especially nitrogen and phosphorus, and waste material cause eutrophication in both fresh and marine ecosystems. Algal blooms from the excess of nutrients decreases light transmission in the water, impacting plants and plankton while reducing the amount of oxygen in the water. As the algae dies, decomposer bacteria consume even more of the dissolved oxygen. In extreme cases, the loss of oxygen results in large dead zones. Runoff of fertilizer and nutrient-rich material from the midwestern United States has caused a 7,728-square-mile oxygen-depleted dead zone in the Gulf of Mexico.

Industrial Wastes in Aquatic Environments

Industrial wastes often pass through the same sewer treatment facilities as domestic wastes. Industrial waste often contains a variety of chemicals and may also contain heavy metals like lead, mercury, cadmium and arsenic. Not all of these chemicals are completely removed in sewage treatment plants, so the chemicals are released into rivers, lakes and marine waters. In addition, some waste may be released or spilled into aquatic ecosystems without any treatment. The effects of sewage pollution on marine life impact organisms throughout the food chain.

Heavy metals build up in fish tissues as the fish consume plankton, algae and smaller prey containing the metals. This process is called biomagnification. As other animals, including humans, eat these fish, the heavy metals can reach sufficient concentrations to poison the consumer. These heavy metals may accumulate in toxic amounts for fish as well.

Control of releases of industrial sewage like petroleum products, radioactive waste and persistent organic pollutants has improved, with oily wastes reduced by 90 percent between the 1980s and 2006. These pollutants caused immediate and long-term effects on ecosystems by poisoning or smothering plankton, plants and animals.

Air Pollution and Aquatic Ecosystems

Industrial soot and smoke also impact aquatic ecosystems. For example, sulfur dioxide combined with water vapor forms sulfuric acid or acid rain. Acid rain and runoff decrease aquatic pH, which interferes with the ability of fish to absorb oxygen, salts and nutrients. A low pH also interferes with calcium absorption. Improper calcium balance for many fish means their eggs do not develop properly, becoming too brittle or weak. Calcium deficiency also causes weak spines and bones in fish and weak exoskeletons for crayfish. Acid rain also leaches aluminum from soils, interfering with reproduction in crustaceans and fish. Furthermore, when pH drops below 6, insects like mayflies and stoneflies can't survive, impacting the food chain.

Litter in Aquatic Ecosystems

Urban sewage includes litter washed into storm drains and eventually into waterways. An estimated 70 percent of this litter ends up on the seabed, about 15 percent lands on beaches and about 15 percent is floating in the ocean. Most of the litter, 70 percent, is plastics with metal and glass making up the majority of the remaining 30 percent. Studies show that more than 1,200 aquatic species interact with the litter by eating it, living in or on it, or getting tangled up in it. Much of the plastic is in the form of microplastics, tiny pieces from the breakdown of larger plastics. Animals as diverse as mammals, fish, crustaceans and others are impacted by this litter.

Effects of Wastewater on the Environment

It's not a secret that there's a real problem in the form of pollution. Many experts estimate that if we are unable to limit global warming to 34.7 degrees Fahrenheit, the consequences may become irreversible. And while it's easy to simply leave the responsibility to the higher powers in society, the indispensable truth is that this requires a collective effort in order to successfully avert this impending disaster.

It's not about drastic change, but rather a gradual and consistent correction of bad habits and the reinforcement of good habits. True change often starts at the base level and works its way up. One of the less covered pollutants comes in the form of wastewater. But before we delve further into its effects on the environment, there's one question that needs to be answered first.

How Bad is Wastewater for the Environment?

The fact that we are making water unusable means that we are literally draining water supplies. Despite the fact that the Earth's surface is mostly made up of bodies of water, consider the fact that we are dumping out waste into those bodies of water. We are literally polluting the water that we could potentially use. But that's not the full extent of the damage that wastewater causes to the environment.

Water and Habitat Contamination

The most immediate effect of wastewater on the environment is when it contributes toward the contamination and destruction of natural habitats and the wildlife that live in those habitats by exposing them to harmful chemicals that would otherwise not be present over the natural course of things.

Squalor

Wastewater is one of the worst sources and carriers of diseases. According to a report from the World Health Organization, more than 3.4 million people die each year from a waterborne disease. Besides the diseases wastewater carries, the combination of human waste, solvents, and paints create fumes that aren't only putrid but they also expose people to harmful fumes. There's a real danger when you inhale sewage gas.

Soil Degradation

Wastewater is often treated and repurposed for use in irrigation. As if that wasn't bad enough on its own, water treatment processes are not completely effective. Chemicals that are harmful to crops may find their way to the soil when the wastewater isn't properly treated. These chemicals will cause the soil to yield fewer crops at a slower rate. Consider also the fact that these crops will eventually be eaten, which can also harm humans.

It Contains Harmful Substances

The composition of wastewater may include heavy metals, pathogens, salts, toxic chemicals, oil and grease, solids, nutrients, sludge, acids and bases, toxic organic compound, organic and inorganic materials. This effluent poses numerous hazards for humans, animals and the environment as a whole. It can be toxic, corrosive, reactive, acidic and ignitable. Therefore, it must undergo treatment before being reused or redirected into the water supply.

Wastewater Effects on Water Bodies

Waterways are generally most at risk to the harmful effects of wastewater. Toxic compounds in the effluent disrupt aquatic ecosystems. When a large amount of biodegradable substances end up in the water, organisms will start to break them down, and they use a lot of dissolved oxygen. Dissolved oxygen is critical for marine life to thrive, and as it becomes depleted, it can be life-threatening for fish.

Wastewater also contains oil and grease that are harder to break down and can settle on the surface of the water. This blocks that light the photosynthetic aquatic plants need. It can also suffocate fish and get caught in birds' feathers. Like heavy metals, such as lead and mercury, these are toxic to both humans and animals. If someone drank from or ate fish from a contaminated water source, they could also suffer from serious health effects. The same happens with the animals and plants living in the water.

Other Harmful Characteristics of Wastewater

When untreated wastewater gets dumped, it's often warm or even hot, which can elevate the temperature of the water, further disrupting the ecosystem. Fish are cold-blooded, so they rely on the water to regulate their own body temperature. If the water is too warm, it can increase respiration, feeding and movement. Additionally, the temperature of the water also affects the amount of oxygen in it.

How do we Mitigate the Effects of Wastewater on the Environment?

As bad as the problem sounds, there are two ways that we can mitigate its impact on the environment. Like most major problems, the wastewater problem needs to be solved through a collective effort. The best way we can contribute towards that effort is to:

Conserve Water

The less water we use, the less wastewater we're going to produce. A few simple ways we can do so is by fixing indoor leaks, reducing shower time, and by avoiding any wasteful washing habits. Switching to water-saving appliances will also help minimize the amount of wastewater produced. This helps water treatment plants process better because if there's less wastewater, the quality of treatment will be better.

Minimize the Polluting Content of Wastewater

The other factor that makes wastewater so harmful lies in the variety of chemicals that mix with human waste and food scraps. This can be averted by choosing the products we buy carefully. A good example of this is by using only low-phosphorus household products in order to reduce phosphate levels in wastewater. Phosphate is a chemical that is found in most home cleaning products.

Innovative Decentralized Water Technology

It requires Intelligence water use as different water uses need different waterQuality. It is required to promote the 3 R approach- Reduce the discharge of WW, Treat or remediate and Reuse. The decentralized technologies are Ecological Sanitation consisting of Septic Tanks, Constructed Wetlands, Composting Toilets, Biodigestor, Anaerobic Filter, Duckweed Lagoons that involve the collection, treatment, and final disposition of the WW on/or close to the location and these are useful in treating wastes from residences, households, small villages, isolated communities, etc.

Advantages of these decentralized systems include the possibility for reusing the effluent and the potentiality for aquifer recharging. A problem in a unit doesn't collapse the whole system. The local potentialities can be developed as small systems can be designed, built and managed by a local professional, improving the local economy.

Wastewater Regeneration Technologies

The technologies control both the aeration supplied to the membranes and the aeration of the biological tanks, using an advanced control system permitting lower air consumption (energy consumption) as well as cutting down on the generation of GH gases (NOx and CO2). It also eliminates emergent pollutants and, when necessary, sludge filterability by adding substances to the activated sludge through automatically controlled dosing.

The final objective will be to create a self-sufficient treatment plant where the energy required for the plant is produced in the plant itself via the obtainment of biogas or other renewable energies. In this regard, the new technologies being developed are aimed at increasing sustainability in the water treatment sector.

References

- Waste-water-pollution: medcraveonline.com, Retrieved 21, April 2020
- Effects-sewage-aquatic-ecosystems-21773: sciencing.com, Retrieved 07, January 2020
- Effects-wastewater-environment: conserve-energy-future.com, Retrieved 12, May 2020

Chapter 4

Sewers and Manholes

A sewer is a large, artificial underground passage or pipe that carries waste and used water away for treatment. Sewer system design plays an important role in city planning because it decides the method of efficient wastewater disposal. Sewer systems are provided with openings, called manholes, through which inspection, maintenance and upgrades can be done. This chapter discusses in detail the different aspects of sewers and manholes.

Sewer

Sewage is water-carried wastes, in either solution or suspension that flow away from a community. It is also known as waste water flows or used water supply of the community. It is more than 99.9% pure water characterized by its volume or rate of flow, its physical condition, its chemical constituents, and the bacteriological organisms that it contains. Depending on their origin, wastewater can be classed as sanitary, commercial, industrial, or surface runoff. The spent water from residences and institutions, carrying body wastes, washing water, food preparation wastes, laundry wastes, and other waste products of normal living, are classed as domestic or sanitary sewage. Liquid-carried wastes from stores and service establishments serving the immediate community, termed commercial wastes, are included in the sanitary or domestic sewage category if their characteristics are similar to household flows.

Wastes that result from an industrial process or the production or manufacture of goods are classed as industrial wastes. Their flows and strengths are usually more varied, intense, and concentrated than those of sanitary sewage. Surface runoff, also known as storm flow or overland flow, is that portion of precipitation that runs rapidly over the ground surface to a defined channel. Precipitation absorbs gases and particulates from the atmosphere, dissolves and leaches materials from vegetation and soil/ suspends matter from the land, washes spills and debris from urban streets and highways, and carries all these pollutants as wastes in its flow to a collection point. Wastewater from all of these sources may carry pathogenic organisms that can transmit disease to humans and other animals; contain organic matter that can cause odor and nuisance problems; hold nutrients that may cause eutrophication of receiving water bodies; and can lead to eco-toxicology. Proper collection and safe, nuisance-free disposal of the liquid wastes of a community are legally recognized as a necessity in an urbanized, industrialized society.

Sewage is created by residences, institutions, hospitals and commercial and industrial establishments. Raw influent (sewage) includes household waste liquid from toilets, baths, showers, kitchens, sinks etc. is disposed of via sewers. In many areas, sewage also includes liquid waste from industry and commerce. The separation and draining of household waste into grey-water and black-water is becoming more common in the developed world, with grey-water being permitted to be used for watering plants or recycled for flushing toilets. A lot of sewage also includes some surface water from roofs or hard-standing areas. Municipal wastewater, therefore, includes residential, commercial, and industrial liquid waste discharges, and may include storm-water run-off. Sewage systems capable of handling storm-water are known as combined systems or combined sewers. Such systems are usually avoided since they complicate and thereby reduce the efficiency of sewage treatment plants owing to their seasonality. The variability in flow also leads to often larger than necessary, and subsequently more expensive, treatment facilities. In addition, heavy storms that contribute more flows than the treatment plant can handle may overwhelm the sewage treatment system, causing a spill or overflow. It is preferable to have a separate storm drain system for storm-water in areas that are developed with sewer systems. As rainfall runs over the surface of roofs and the ground, it may pick up various contaminants, including soil particles and other sediment, heavy metals, organic compounds, animal waste, and oil and grease. Some jurisdictions require storm-water to receive some level of treatment before being discharged directly into waterways. Examples of treatment processes used for storm-water include sedimentation basins, wetlands, buried concrete vaults with various kinds of filters, and vortex separators to remove coarse solids).

Sewage Services

This involves collection of sewage, treatment and disposal.

Collection of Sewage

Sewage collection and disposal systems transport sewage through cities and other inhabited areas to sewage treatment plants in order to protect public health and prevent disease. Sewage is treated in order to control water pollution before discharge to surface waters. A sewage system may convey the wastewater by gravity to a sewage treatment plant. Where pipeline excavation is difficult because of rock or there is limited topographic relief (i.e., due to flat terrain), gravity collection systems may not be practical and the sewage must be pumped through a pipeline to the treatment plant. In low-lying communities, wastewater may be conveyed by vacuum. Pipelines range in size from pipes of six inches (150mm) in diameter to concrete-lined tunnels of up to thirty feet (10m) in diameter.

Sewage can also be collected by low pressure pumps and vacuum systems. A low pressure system uses a small grinder pump located at each point of connection, typically a house or business. Vacuum sewer systems use differential atmospheric pressure to move the liquid to a central vacuum station. Typically, a vacuum sewer station can service approximately 1,200 homes before it becomes more cost-effective to build another station. A system of sewer pipes (sewers) collects sewage and takes it for treatment or disposal. The system of sewers is called sewerage or sewerage system. Where a main sewerage system has not been provided, sewage may be collected from homes by pipes into septic tanks or cesspits, where it may be treated or collected in vehicles and taken for treatment or disposal. Properly functioning septic tanks require emptying every 2-5 years depending, on the load of the system.

Historical Sewage Conveyance

The historical focus of sewage treatment was on conveyance of raw sewage to a natural body of water, such as a river or ocean, where it would be satisfactorily diluted and dissipated. Early human habitations were often built next to water sources. Rivers could double as a crude form of natural sewage disposal. The Indus architects designed sewage disposal systems on a large scale, building networks of brick effluent drains following the lines of the streets. The drains were seven to ten feet wide, cut at two feet below ground level with U-shaped bottoms lined with loose brick easily taken up for cleaning. At the intersection of two drains, the sewage planners installed cesspools with steps leading down into them, for periodic cleaning. By -^2700 B.C., these cities had standardized earthenware plumbing pipes with broad flanges for easy joining with asphalt to stop leaks.

Ancient Systems

The first sanitation system has been found at the prehistoric Middle East and the surrounding areas. The first time an inverted siphon system was used, along with glass covered clay pipes, was in the palaces of Crete, Greece. It is still in working condition, after about 3000 years. Ancient Minoan civilization had stone sewers that were periodically flushed with clean water. Roman towns and garrisons in the United Kingdom between 46 BC and 400 CE had complex sewer networks sometimes constructed out of hollowed out Elm logs which were shaped so that they butted together with the down-stream pipe providing a socket for the upstream pip.

Higher population densities required more complex sewer collection and conveyance systems in order to maintain (somewhat) sanitary conditions in crowded cities. The ancient cities of Harappa and Mohenjo-daro of the Indus Valley civilization constructed complex networks of brick-lined sewage drains from around 2.600 BC and also had outdoor flush toilets connected to this network. The urban areas of the Indus Valley civilization provided public and private baths. Sewage was disposed through underground drains built with precisely laid bricks, and a sophisticated water management system with numerous reservoirs was established. In the drainage systems, drains from houses were connected to wider public drains. A significant development was the construction of a network of sewers to collect waste water, which began from the Indus Valley civilization. In some cities, including Rome and Istanbul (Constantinople), networked ancient sewer systems continue to function today as collection systems for those cities' modernized sewer systems. Instead of flowing to a river or the sea, the pipes have been re-routed to modern sewer treatment facilities.

The system remained without much progress until the 16th century, where, in England, Sir John Harington invented a device for Queen Elizabeth (his godmother) that released wastes into cesspools. However, many cities had no sewers and relied on nearby rivers or occasional rain to wash away sewage. In some cities, waste water simply ran down the streets, which had stepping stones to keep pedestrians out of the muck, and eventually drained as run-off into the local watershed. This was enough in early cities with few occupants, but the growth of cities quickly over-polluted streets and became a constant source of disease. Even as recently as the late 19th century, sewerage systems in parts of the highly industrialized United Kingdom were reported to be so inadequate that -water-borne diseases such as cholera and typhoid were still common. In Merthyr Tydfil, a large town in South Wales, most houses discharged their sewage to individual cess-pits which persistently overflowed, causing the pavements to be awash with foul sewage.

Industrial Revolution Era

As an outgrowth of the Industrial Revolution, many cities in Europe and North America grew in the 19th century, frequently leading to crowding and increasing concerns about public health. As part of a trend of municipal sanitation programs in the late 19th and 20th centuries, many cities constructed extensive sewer systems to help control outbreaks of disease. Initially, these systems discharged sewage directly to surface waters without treatment. As pollution, of water bodies became a concern, cities added sewage treatment plants to their systems.

Design and Analysis of Collection Systems

Design and sizing of sewage collection systems considers population served, commercial and industrial flows, flow peaking characteristics and wet weather flows. Combined sewer systems are designed to transport both storrn water run-off and sewage in the same pipe. Besides the projected sewage flow, the size and characteristics of the watershed are the overriding design considerations for combined sewers. Often, combined sewers cannot handle the volume of run-off, resulting in combined sewer overflows and causing water pollution problems in nearby water bodies.

Separate sanitary sewer systems are designed to transport sewage alone. In communities served by separate sanitary sewers, another pipe system is constructed to convey storm water run-off directly to surface waters. Most municipal sewer systems constructed today are separate sewer systems. Although separate sewer systems are intended to transport only sewage, all sewer systems have some degree of inflow and infiltration of surface water and groundwater, which can lead to sanitary sewer overflows. Inflow and infiltration is highly affected by antecedent moisture conditions, which also represents an important design consideration in these systems. A sewer bed is a piece of land typically used by a municipality for the dumping of raw sewage. Usually, raw sewage was brought by truck or drawn by horses to be dumped, but the practice stopped back in the 1940s.

Sewage Treatment

The objective of sewage treatment is to produce a disposable effluent without causing harm or trouble to the communities and prevent pollution. Sewage treatment, or domestic wastewater treatment, is the process of removing contaminants from wastewater and household sewage, both run-off (effluents) and domestic, to produce liquid and solid (sludge) suitable for discharge to the environment or for reuse. It is a unit process used to separate, modify, remove, and destroy objectionable, hazardous, and pathogenic substances carried by wastewater in solution or suspension in order to render the water fit and safe for intended uses. It is a form of waste management. A septic tank or other on-site wastewater treatment system such as bio-filters can be used to treat sewage close to where it is created. It includes physical, chemical, and biological processes to remove physical, chemical and biological contaminants. Its objective is to produce a waste stream (or treated effluent) and a solid waste or sludge suitable for discharge or re-use back into the environment. This material is often inadvertently contaminated with many toxic organic and inorganic compounds.

Sewage water is a complex matrix, with many distinctive chemical characteristics. These include high concentrations of ammonium, nitrate, phosphorus, high conductivity (due to high dissolved solids), high alkalinity, with pH typically ranging between 7 and 8. Tri-halo-methanes are also likely to be present as a result of past disinfection. In developed countries sewage collection and treatment is typically subject to local, state and federal standards. Treatment removes unwanted constituents without affecting or altering the water molecules themselves, so that wastewater containing contaminants can be converted to safe drinking water. Stringent water quality and effluent standards have been developed that require reduction of suspended solids (turbidity), biochemical oxygen demand (related to degradable organics), and coliform organisms (indicators of fecal pollution); control of pH as well as the concentration of certain organic chemicals and heavy metals: and use of bioassays to guarantee safety of treated discharges to the environment. In all cases, the impurities, contaminants, and solids removed from all wastewater treatment processes must ultimately be collected, handled, and disposed off safely, without damage to humans or the environment.

Treatment processes are chosen on the basis of composition, characteristics, and concentration of materials present in solution or suspension. The processes are classified as pretreatrnent, preliminary, primary, secondary, or tertiary treatment, depending on type, sequence, and method of removal of the harmful and unacceptable constituents. Pretreatrnent processes equalize flows and loadings and precondition wastewaters to neutralize or remove toxics and industrial wastes that could adversely affect sewers or inhibit operations of publicly owned treatment works. Preliminary treatment processes protect plant mechanical equipment; remove extraneous matter such as grit, trash, and debris; reduce odors; and render incoming sewage more amenable to subsequent treatment and handling.

Primary treatment employs mechanical and physical unit processes to separate and remove floatables and suspended solids and to prepare wastewater for biological treatment. Secondary treatment utilizes aerobic microorganisms in biological reactors to feed on dissolved and colloidal organic matter. As these microorganisms reduce biochemical oxygen demand and turbidity (suspended solids), they grow, multiply, and form an organic floe, which must be captured and removed in final settling tanks.

Tertiary treatment, or advanced treatment, removes specific residual substances trace-= organic materials, nutrients, and other constituents that are not removed by biological processes. Most advanced wastewater treatment systems include de-nitrification and ammonia stripping, carbon adsorption of trace organics, and chemical precipitation. Evaporation, distillation, electro-dialysis, ultra-filtration, reverse osmosis, freeze drying, freeze-thaw, floatation, and land application, with particular emphasis on the increased use of natural and constructed wetlands, are being studied and utilized as methods for advanced wastewater treatment to improve the quality of the treated discharge to reduce unwanted effects on the receiving environment- On-site sewage treatment for individual homes or small institutions uses septic tanks, which provide separation of solids in a closed, buried unit. Effluent is discharged to subsurface absorption systems.

Process

Sewage can be treated-close to where it is created (in septic tanks, bio-filters or aerobic treatment systems), or collected and transported via a network of pipes and pump stations to a municipal treatment plant. Sewage collection and treatment is typically subject to local, state and federal regulations and standards. Industrial sources of wastewater often require specialized treatment processes. Conventional sewage treatment may involve throe stages, called primary, secondary and tertiary treatment.

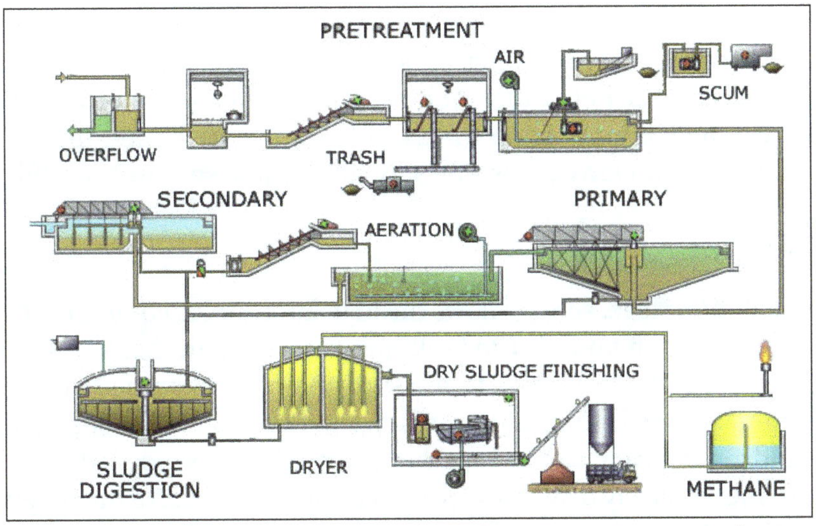

Figure: Process flow diagram for a typical large-scape treatment plant.

Primary treatment consists of temporarily holding the sewage in a quiescent basin where heavy solids can settle to the bottom, while oil, grease and lighter solids float to the surface. The settled and floating materials are removed and the remaining liquid may be discharged or subjected to secondary treatment. Secondary treatment removes dissolved and suspended biological matter.

Secondary treatment is typically performed by indigenous, water-borne micro-organisms in a managed habitat. Secondary treatment may require a separation process to remove the micro-organisms from the treated water prior to discharge or tertiary treatment.

Tertiary treatment is sometimes defined as anything more than primary and secondary treatment. Treated water is sometimes disinfected chemically or physically (for example by lagoons and micro-filtration) prior lo discharge into a stream, river, bay, lagoon or wetland, or it can be used for the irrigation of a golf course, green way or park. If it is sufficiently clean, it can also be used for groundwater recharge or agricultural purposes.

Pre-Treatment

Pre-treatment removes materials that can be easily collected from the raw wastewater before they damage or clog the pumps and skimmers of primary treatment clarifiers.

Screening

The influent sewage water is strained to remove all large objects carried in the sewage stream, such as rags, sticks, tampons, cans, fruit, etc. This is most commonly done with an automated mechanically raked bar screen in modern plants serving large populations, whilst in smaller or less modern plants manually cleaned screen may be used. The raking action of a mechanical bar screen is typically paced according to the accumulation on the bar screens and/or flow rate. The solids are collected and later disposed in a landfill or incinerated.

Grit Removal

Pre-treatment may include a sand or grit channel or chamber (sometimes called a de-gritter) where the velocity of the incoming wastewater is carefully controlled to allow sand, grit and stones to settle, while keeping the majority of the suspended organic material in the water column. Sometimes there is a sand washer (grit classifier) followed by a conveyor that transports the sand to a container for disposal. The contents from the sand catcher may be fed into the incinerator in a sludge processing plant, but in many cases, the sand and grit is sent to a landfill.

Primary Treatment

In the primary sedimentation stage, sewage flows through large tanks, commonly called "primary clarifiers" or "primary sedimentation tanks". The tanks are large enough that sludge can settle and floating material such as grease and oils can rise to the surface and be skimmed off. The main purpose of the primary sedimentation stage is to produce both a generally homogeneous liquid capable of being treated biologically and a sludge that can be separately treated or processed. Primary settling tanks are usually equipped with mechanically driven scrapers that continually drive the collected sludge towards a hopper in the base of the tank from where it can be pumped to further sludge treatment stages.

Secondary Treatment

Secondary treatment is designed to substantially degrade the biological content of the sewage such as are derived from human waste, food waste, soaps and detergent. The majority of municipal plants treat the settled sewage liquor using aerobic biological processes. For this to be effective, the biota requires both oxygen and a substrate on which to live. There are a number of ways in which this is done. In all these methods, the bacteria and protozoa consume biodegradable soluble organic contaminants (e.g. sugars, fats, organic short-chain carbon molecules, etc.) and bind much of the less soluble fractions into floe. Secondary treatment systems are classified as fixed-film or suspended-growth.

Fixed-film or attached growth system treatment process including trickling filter and rotating biological contactors where the biornass grows on media and the sewage passes over its surface. In suspended-growth systems, such as activated sludge, the biomass is well mixed with the sewage and can be operated in a smaller space than fixed-film systems that treat the same amount of water. However, fixed-film systems are more able to cope with drastic changes in the amount of biological material and can provide higher removal rates for organic material and suspended solids than suspended growth systems. Roughing filters are intended to treat particularly strong or variable organic loads, typically industrial, to allow them to then be treated by conventional secondary treatment processes. Characteristics include typically tall, circular filters filled with open synthetic filter media to which wastewater is applied at a relatively high rate. They are designed to allow high hydraulic loading and a high flow-through of air. On larger installations, air is forced through the media using blowers. The resultant wastewater is usually within the normal range for conventional treatment processes.

Activated Sludge

It is a process dealing with the treatment of sewage and industrial waters. In general, activated sludge plants encompass a variety of mechanisms and processes that use dissolved oxygen to promote the growth of biological floe that substantially removes organic material. The process traps particulate material and can, under ideal conditions, convert ammonia to nitrite and nitrate and ultimately to nitrogen gas. Most biological oxidation processes for treating industrial wastewaters have in common the use of oxygen (or air) and microbial action. Surface-aerated basins achieve 80-to 90% removal of

Biochemical Oxygen Demand with retention times of 1 to 10 days, and the basins may range in depth from 1.5 to 5.0 meters and use motor-driven aerators floating on the surface of the wastewater. In an aerated basin system, the aerators provide two functions: they transfer air into the basins required by the biological oxidation reactions, and they provide the mixing required for dispersing the air and for contacting the reactants (that is, oxygen wastewater and microbes).Typically, the floating surface aerators are rated to deliver the amount of air equivalent to 1.8 to 2.7kg O_2/kW-h. However, they do not provide as good mixing as is normally achieved in activated sludge systems; hence, aerated basins do not achieve the same performance level as activated sludge units. Biological oxidation processes are sensitive to temperature and, between 0°C and 40°C, the rate of biological reactions increase with temperature. Most surface aerated vessels operate at between 4°C and 32°C.

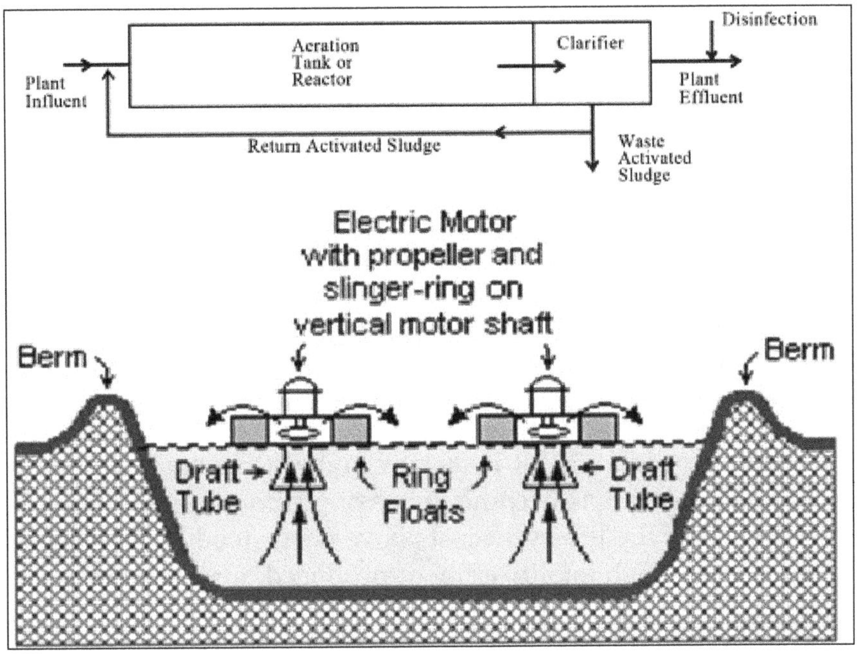

Figure: A typical surface-aerated basin (using motor driven floating aerators).

Trickling Filter

In older plants and plants receiving more variable loads, trickling filter beds are used where the settled sewage liquor is spread onto the surface of a deep bed made up of coke (carbonized coal), limestone chips or specially fabricated plastic media. Such media must have high surface area to support the biofilms that form. The liquor is distributed through perforated rotating arms radiating from a central pivot. The distributed liquor trickles through this bed and is collected in drains at the base. These drains also provide a source of air which percolates up through the bed, keeping it aerobic. Biological films of bacteria, protozoa and fungi form on the media surface and digest or otherwise reduce the organic content. This biofilm is grazed by insect larvae and worms which help maintain an optimal thickness. Overloading of beds increases the thickness of the film, leading to clogging of the filter media and ponding on the surface.

Biological Aerated Filters

Biological Aerated (or Anoxic) Filter (BAF) or Bio-filters combine filtration with biological carbon reduction, nitrification or de-nitrification; BAF usually includes a reactor filled with a filter media. The media is either in suspension or supported by a gravel layer at the foot of the filter. The dual purpose of this media is to support highly active biomass that is attached to it and to filter suspended solids. Carbon reduction and ammonia conversion occurs in aerobic mode, and sometimes achieved in a single reactor, while nitrate conversion occurs in anoxic mode. BAF is operated either in up-flow or down-flow configuration depending on design specified by manufacturer.

Membrane Bioreactors

Membrane bioreactors (MBR) combine activated sludge treatment with a membrane liquid-solid separation process. The membrane component uses low pressure micro-filtration or ultra-filtration membranes and eliminates the need for clarification and tertiary filtration. The membranes are typically immersed in the aeration tank (however, some applications utilize a separate membrane tank). One of the key benefits of a membrane bioreactor system is that it effectively overcomes the limitations associated with poor settling of sludge in conventional activated sludge (CAS) processes. The technology permits bioreactor operation with considerably higher mixed liquor suspended solids (MLSS) concentration than CAS systems, which are limited by sludge settling. The process is typically operated at MLSS in the range of 8,000-12,000 mg/L, while CAS are operated in the range of 2,000-3,000 mg/L. The elevated biomass concentration in the membrane bioreactor process allows for very effective removal of both soluble and particulate biodegradable materials at higher loading rates. Increased Sludge Retention Times (SRTs) — usually exceeding 15 days—ensure complete nitrification even in extremely cold weather.

The cost of building and operating a MBR is usually higher than conventional wastewater treatment. However, as the technology has become increasingly popular and has gained wider acceptance throughout the industry, the life-cycle costs have been steadily decreasing. The small footprint of MBR systems, and the high quality effluent produced, makes them particularly useful for water reuse applications.

Secondary Sedimentation

The final step in the secondary treatment stage is to settle out the biological floe or filter material and produce sewage water containing very low levels of organic material and suspended matter.

Rotating Biological Contactors

Rotating biological contactors (RBCs) are mechanical secondary treatment systems which are robust and capable of withstanding surges in organic load. The rotating disks support the growth of bacteria and microorganisms present in the sewage, which breakdown and stabilize organic pollutants. To be successful, microorganisms need both oxygen to live and food to grow. Oxygen is obtained from the atmosphere as the disks rotate. As the micro-organisms grow, they build up on the media until they are sloughed off due to shear forces provided by the rotating discs in the sewage. Effluent from the RBC is then passed through final clarifiers where the microorganisms in suspension settle as sludge. The sludge is withdrawn from the clarifier for further treatment.

A functionally similar biological filtering system has become popular as part of home aquarium filtration and purification. The aquarium water is drawn up out of the tank and then cascaded over a freely spinning corrugated fiber-mesh wheel before passing through a media filter and back into the aquarium. The spinning mesh-wheel develops a bio-film coating of microorganisms that feed on the suspended wastes in the aquarium water and are also exposed to the atmosphere as the wheel rotates. This is especially good at removing waste urea and ammonia urinated into the aquarium water by the fish and other animals.

Tertiary Treatment

The purpose of tertiary treatment is to provide a final treatment stage to raise the effluent quality before it is discharged to the receiving environment (sea, river, lake, ground, etc.). More than one tertiary treatment process may be used at any treatment plant. If disinfection is practiced, it is always the final process. It is also called "effluent polishing "CAS systems, which are limited by sludge settling. The process is typically operated at MLSS in the range of 8,000-12,000 mg/L, while CAS is operated in the range of 2,000-3,000 mg/L the elevated biomass concentration in the membrane bioreactor process allows for very effective removal of both soluble and particulate biodegradable materials at higher loading rates. Increased Sludge Retention Times (SRTs)—usually exceeding 15 days— ensure complete nitrification even in extremely cold weather.

Filtration

Sand filtration removes much of the residual suspended matter. Filtration over activated carbon removes residual toxins.

Lagooning

Lagooning provides settlement and further biological improvement through storage in large man-made ponds or lagoons. These lagoons are highly aerobic and colonization by native macrophytes, especially reeds, is often encouraged. Small filter feeding invertebrates such as Daphnia and species of Rotifera greatly assist in treatment by removing fine particulates.

Constructed Wetlands

Constructed wetlands include engineered reed-beds and a range of similar methodologies, all of which provide a high degree of aerobic biological improvement and can often be used instead of secondary treatment, for small communities.

Nutrient Removal

Water nitrification itself is a two-step aerobic process, each step facilitated by a different type of bacteria. The oxidation of ammonia (NH_3) to nitrite (NO_2-) is most often facilitated by Nitrosomonas species (nitroso referring to the formation of a nitroso functional group). Nitrite oxidation to nitrate (NO3-), though traditionally believed to be facilitated by Nitrobacter spp. (nitro referring to the formation of a nitro functional group}, is now known to be facilitated in the environment almost exclusively by Nitrospira spp. De-nitrification requires anoxic conditions to encourage the appropriate biological communities to form. It is facilitated by a wide diversity of bacteria. Sand filters, lagooning and reed-beds can all be used to reduce nitrogen, but the activated sludge process (if designed well) can do the job the most easily. Since de-nitrificalion is the reduction of nitrate to di-nitrogcn gas, an electron donor is needed, and can be done, depending on the wastewater, organic matter (from faeces), sulfide, or an added donor like methanol. Sometimes the conversion of toxic ammonia to nitrate alone is referred to as tertiary treatment. Many sewage treatment plants use axial flow pumps to transfer the nitrified mixed liquorfrom the aeration zone to the anoxic zone for de-nitrification. These pumps are often referred to as Internal Mixed Liquor Recycle pumps (IMLR pumps).

Phosphorus Removal

Phosphorus removal is important as it is a limiting nutrient for algal growth in many fresh water systems. It is also particularly important for water re-use systems where high phosphorus concentrations may lead to fouling of downstream equipment such as reverse osmosis. Phosphorus can be removed biologically in a process called enhanced biological phosphorus removal. In this process, specific bacteria called polyphosphate accumulating organisms (PAOs), are selectively enriched and accumulate large quantities of phosphorus within their cells (up to 20% of their mass). When the biornass enriched in these bacteria is separated from the treated water, these bio-solids have a high fertilizer value. Phosphorus removal can also be achieved by chemical precipitation, usually with salts of iron (e.g. ferric chloride), aluminum (e.g. alum), or lime. This may lead to excessive sludge productions as hydroxides precipitate and the added chemicals can be expensive. Chemical phosphorus removal requires significantly smaller equipment footprint than biological removal is easier to operate, and is often more reliable than biological phosphorus removal. Once removed, phosphorus, in the form of a phosphate rich sludge, may be stored in a land filled or resold for use in fertilizer production.

Disinfection

The purpose of disinfection in the treatment of wastewater is to substantially reduce the number of microorganisms in the water to be discharged back into the environment. The effectiveness of disinfection depends on the quality of the water being treated (e.g., cloudiness, pH, etc.}, the type of disinfection being used, the disinfectant dosage (concentration and time), and other environmental variables. Cloudy water will be treated less successfully since solid matter can shield organisms, especially, from ultraviolet light or if contact times are low. Generally, short contact times, low doses and high flows militate against effective disinfection. Common methods of disinfection include ozone, chlorine, or ultraviolet light. Chloramine, which is used for drinking water, is not used in wastewater treatment because of its persistence. Chlorination remains the most common form of wastewater disinfection in North America and in Nigeria due to its low cost and long-term history of effectiveness. One disadvantage is that chlorination of residual organic material can generate chlorinated-organic compounds that may be carcinogenic or harmful to the environment. Residual chlorine or chloramines may also be capable of chlorinating organic material in the natural aquatic environment. Further, because residual chlorine is toxic to aquatic species, the treated effluent must also be chemically de-chlorinated, adding to the complexity and cost of treatment.

Ultraviolet (UV) light can be used instead of chlorine, iodine, or other chemicals. As no chemicals are used, the treated water has no adverse effect on organisms that later consume it, as may be the case with other methods. UV radiation causes damage to the genetic structure of bacteria, viruses, and other pathogens, making them incapable of reproduction. The key disadvantages of UV disinfection are the need for frequent lamp maintenance and replacement, and the need for a highly treated effluent to ensure that the target microorganisms are not shielded from the UV radiation (i.e., any solids present in the treated effluent may protect microorganisms from the UV light). Ozone O_3 is generated by passing oxygen O_2 through a high voltage potential, resulting in a third oxygen atom becoming attached and forming O_3. Ozone is very unstable and reactive, and oxidizes most organic materials it comes in contact with, thereby destroying many pathogenic microorganisms. Ozone is considered to be safer than chlorine because, unlike chlorine which has to be stored on site (highly poisonous in the event of an accidental release), ozone is generated onsitc as needed. Ozonation also produces fewer disinfection byproducts than chlorination. A disadvantage of ozone disinfection is the high cost of the ozone generation equipment and the requirements for special operators.

Package Plants and Batch Reactors

Package plants may be-referred to as high charged or low charged. This refers to the way the biological load is processed. In high charged systems, the biological stage is presented with a high organic load and the combined floe and organic material is then oxygenated for a few hours before being charged again with a new load. In the low charged system the biological stage contains a low organic load and is combined with flocculate for a relatively long time. In order to use less space, treat difficult waste, deal with intermittent flow or achieve higher environmental standards, a number of designs of hybrid treatment plants have been produced. Such plants often combine all or at least two stages of the three main treatment stages into one combined stage. One type of system that combines secondary treatment and settlement is the sequencing batch reactor (SBR). Typically, activated sludge is mixed with raw incoming sewage and mixed and aerated. The resultant mixture is then allowed to settle producing a high quality effluent. The settled sludge is run off and re-aerated before a proportion is returned to the head-works. The disadvantage of such processes is that precise control of timing, mixing and aeration is required.

Types of Sewer

- Asbestos Cement (AC) Sewer.
- Brick Sewer.
- Cement Sewer.
- Cast Iron (CT) Sewer.
- Steel Sewers.
- Plastic Sewers.

Asbestos Cement (AC) Sewer

Types of sewer like Asbestos Cement (AC) Sewers are manufactured from a mixture of cement and asbestos fiber. Asbestos Cement (AC) Sewers are suitable for carrying domestic sanitary sewage. Asbestos cement sewer is best as vertical pipe for carrying sullage from upper floors of multistory buildings (in two pipe system of plumbing).

Advantages of Asbestos Cement (AC) Sewer

- Smooth.
- Light in weight.

- Can easily be cut, fitted and drilled.
- Durable against soil corrosion.

Disadvantages of Asbestos Cement (AC) Sewer

- Brittle cannot withstand heavy loads.
- They are easily broken in handling and transport.

Brick Sewers

These types of sewer (Brick Sewers) are made at site and used for construction large size sewer. Brick Sewers are very useful for construction of storm sewer or combined sewer. Nowadays brick sewers are replaced by concrete sewer. Brick sewers my get deformed and leakage may take place. A lot of labour work is required. To avoid leakage the brick sewer should be plastered.

Cement Concrete

- PCC - for dia upto 60 cm.

Suitable for small storm drains. Not durable.

- RCC - for dia > 60 cm.

They may be cast in situ or precast, resistant to heavy loads, corrosion and high pressure. These are very heavy and difficult to transport.

Cast Iron (CI) Sewers

These types of sewer are High strength and durability water tight. Cast Iron sewers can withstand high internal pressure and can bear external load. Cast Iron sewers are suitable for the following conditions:

- When the sewage is conveyed under high pressure.
- When the sewer line is subject to heavy external load e.g. under railway line, foundation wall etc, below highways.
- When there is considerable difference in temperature.

Steel Sewers

These types of sewer (steel sewers) are Impervious; light, resistant to high pressure, flexible, suitable when:

- The sewage is carried under pressure.
- The sewage has to be carried across a river under water.
- The sewer has to cross under a railway track.
- They are generally used for outfall and trunk sewers.

Plastic Sewers

Nowadays PVC sewers are used for carrying sewage. Plastic sewers are resistant to corrosion. Such types of sewer are light in weight, smooth and can be bent easily. But these types of sewer (Plastic sewers) are having high co-efficient of thermal expansion and cannot be used in very hot areas.

Other Types of Sewer Materials

- Wooden Sewers (Rare now).
- Stoneware Sewers.

Sewer System Design

Cost Considerations

To arrive at the objective function to be minimized, it is necessary to describe the cost function of the various components of a sewer line. Equation. shows the ith link of a sewer line having nodes i -1 and i. The depths at nodes i - 1 and i are and d_{i-1} d_i, respectively. The link has a trench bottom width w_i, sewer diameter D_i, and length L_i. The cost of the ith link of sewer pipe C_{pi} was expressed:

$$C_{pi} = k_m L_i D_i^m$$

where k_m and m = pipe cost parameters, which can be obtained by plotting the per meter cost of pipes of various diameters on a double logarithmic paper. The ith link of the sewer trench has a trapezoidal cross section having a side slope 0.1 horizontal to 1 vertical. As an approximation, assuming vertical sides, the earthwork cost of the ith link C_{ei} was written:

$$C_{ei} = \frac{1}{2} L_i w_i \left[c_e(d_{i-1} + d_i) + \frac{1}{3} c_r (d_{i-1}^2 + d_{i-1} d_i = d_i^2 \right] + c_s L_i (d_{i-1} + d_i)$$

where c_e = unit excavation cost at ground level [monetary unit per unit volume of earthwork (dollars/m^{-3})]; c_r = increase in unit excavation cost per unit depth [monetary unit per unit volume per unit depth (dollars/m^{-4}); and c_s = cost of sheeting and shoring per unit area [monetary unit per unit area (dollars/ m^{-2})]. have studied the excavation costs of various types of strata and found that c_e/c_r varied between 6.96 and 13.86 m. According to Manual, wi follows the following criteria:

$$w_i = 1, D_i \le 0.6\,m$$
$$w_i = D_i + 0.4, \ D_i > 0.6\,m$$

Analysis of manhole costs for various depths and sizes of sewers revealed that for D ≤0.6 m, the cost of a manhole is an insensitive function of D and it solely depends on the depth of the manhole. The cost analysis expresses the cost of ith manhole C_{hi}:

$$C_{hi} = k_h d_i$$

where k_h = manhole cost coefficient. For larger depths, k_h increases with the Di. Using the cost function F of the sewer line having n links was written:

$$F = k_h d_0 + \sum_{i=1}^{n}\left(L_i \{k_m D_i^m + \frac{1}{2} w_i \left[c_e (d_{i-1} + d_i) + \frac{1}{3} c_r (d_{i-1}^2 + d_{i-1} d_i + d_i^2\right] + c_s (d_{i-1} + d_i)\} + k_h d_i \right)$$

where d_o = initial nodal depth. The earthwork cost as given by can be rewritten in the following form:

$$C_{ei} = k_{ei} Li(d_{i-1} + d_i)$$

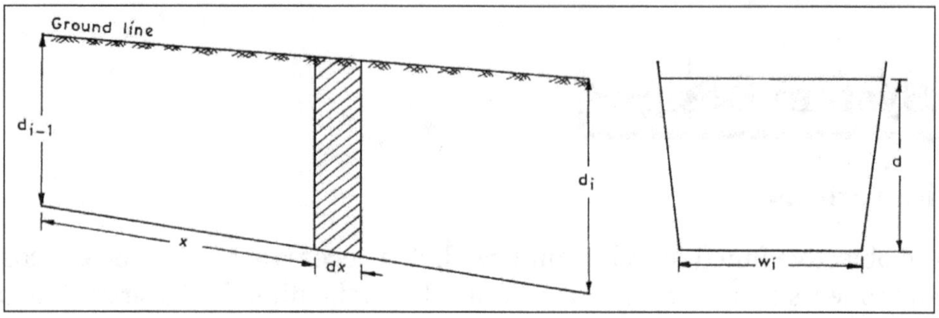

Figure: Definition Sketch: Sewer Link.

where k_{ei} = earthwork cost coefficient given by:

$$k_{ei} = \frac{1}{2} w_i \left(c_e + \frac{1}{3} c_r \frac{d_{i-1}^2 + d_{i-1} d_i + d_i^2}{d_{i-1} + d_i} \right) + c_s$$

Using, the cost function was given by:

$$F = (k_h + k_{e1} L_1) d_0 + k_{en} L_n + k_h) d_n + \sum_{i=1}^{n} k_m L_i D_i^m$$

$$+ \sum_{i=1}^{n-1} (k_{ei} L_i + k_{ei+1} L_{i+1} + k_h) d_i$$

where dn = terminal nodal depth.

Resistance Relationships

Resistance Equation

The following dimensionally consistent resistance equation for uniform free surface flow:

$$V = -2.457 \sqrt{gRS_0}\, 1 1n(\frac{\varepsilon}{12R} + \frac{0.221v}{R\sqrt{gRS_o}}$$

where V = average flow velocity; g = gravitational acceleration; R = hydraulic radius defined as the ratio of flow area A to wetted perimeter P; ε = average roughness height; n = kinematic viscosity of water; and S_0 = invert slope. Using and the definition of discharge Q = AV, the discharge Q was written:

$$Q = -2.457 A\sqrt{gRS_0}\, 1 1n\left(\frac{\varepsilon}{12R} + \frac{0.221v}{R\sqrt{gRS_o}}\right.$$

Maximum Velocity and Discharge Equations

For a circular section A and P are given by:

$$A = 0.25D^2[\cos^{-1}(1-2\eta) - 2(1-2\eta)\sqrt{\eta(1-\eta)}]$$

$$P = D\cos^{-1}(1-2\eta)$$

where η = relative flow depth defined as the ratio of the normal depth (depth of uniform flow) to sewer diameter D. Using and maximizing V, it was found that the maximum average velocity V_{max} occurs at the flow depth 0.813D. For this flow depth and using, V_{max} was obtained:

$$V_{max} = -1.355\sqrt{gDS_0}\; 1n\left(\frac{\varepsilon}{3.65D} + \frac{1.32v}{D\sqrt{gDS_0}}\right)$$

Maximizing Q in, it was found that the maximum dis charge Q_{max} occurs at the flow depth 0.943D, for which gave Q_{max}:

$$Q_{max} = -D^2\sqrt{gDS_0}\; 1n\left(\frac{\varepsilon}{3.65D} + \frac{1.43v}{D\sqrt{gDS_0}}\right)$$

Diameter Equation

Using and adopting a procedure similar to the pipe diameter equation obtained by the following equation for sewer diameter under partly full flow condition was obtained:

$$D = k_D\left[\varepsilon^{1.25}\left(\frac{Q^2}{gS_0}\right)^{4.75} + \frac{vQ^{9.4}}{(gS_0)^{5.2}}\right]^{0.04}$$

where k_D = diameter coefficient given by:

$$k_D = \frac{0.442}{\eta^{0.8}} + 0.164\eta + \left(\frac{\eta}{1.2}\right)^{16}$$

The maximum error in the use of is <0.5%.

Normal Depth Equation

Using, the following equation was fitted for nondimensional normal depth:

$$\eta = 0.943(0.0683\{\left(\frac{Q_{max}}{Q}\right)^{0.886}\left[1+\left(\frac{Q_{max}}{Q}\right)^{1.69}\right] - 2\}^{0.357} + 1)^{-12.25}$$

where Q_{max} is obtained from. Barring small depths $\eta \leq 0.3$, the maximum error involved in is 1.5%.

Constraints Description

Outfall Depth Constraint

The invert slope of ith link S_{oi} was given by:

$$S_{0i=}\frac{d1 - d_{i-1} + z_{i-1} - z_i}{L_i}$$

where z_i = ground elevation at ith node. Eq. was rewritten:

$$d_i = L_i S_{0i} + d_{i-1} + z_i - z_{i-1}$$

Starting from i = 1, repeated use of yielded:

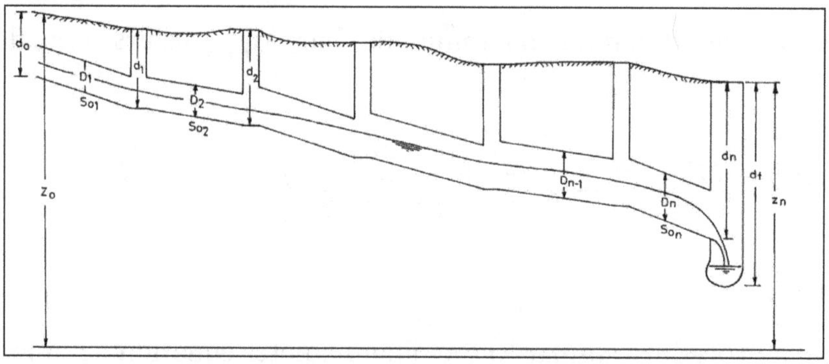

Figure: Definition Sketch: Longitudinal Profile.

$$d_i = d_0 + z_i - z_0 + \sum_{j-1}^{i} L_j S_{0j}$$

where the initial nodal depth do was obtained by adopting the minimum cover h_{min}. Thus:

$$d_0 = D_1 + h_{min}$$

It was found that the rounding off of D_1 changes do in the bits of 0.05 m. Thus there is considerable change in the slope S_{01} as revised by using. For avoiding this sensitivity in S_{01}, it is sufficient to fix d_o:

$$d_0 = h_{min} + 0.5m$$

The terminal nodal depth $d_n \leq$ outfall depth d_f. For a branch sewer line, d_f is provided by the depth of the receiving sewer line at the outfall nodal point. On the other hand, for a main sewer line, d_f is fixed by topographic conditions. Thus, using with i = n, the condition translates to:

$$d_0 + z_n - z_o + \sum_{j=1}^{n} L_j S_{0j} \leq d_f$$

Flow Constraint

The roughness heights for various materials have been given in books of fluid mechanics. As ε increases with time, the largest possible value of ε should be adopted in. Thus simplifying for the ith link, by dropping the term containing v and rearranging the terms, the following equation was obtained:

$$S_{0i} = \left(\frac{D_{si}}{D_i}\right)^{5.26}$$

$$D_{si} = k_{Di}\varepsilon^{0.05}\left(\frac{Q_i^2}{g}\right)^{0.19}$$

where k_{Di} = diameter coefficient for ith link, which can be obtained and Table. Using, the constraints and respectively were written:

$$d_i = d_0 + z_i - z_0 + \sum_{j=1}^{i} L_j \left(\frac{D_{sj}}{D_j}\right)^{5.26}$$

Table: Relative Depth:

Sewer diameter D (m)	Relative depth η
0.15–0.25	0.50
0.30–0.50	0.60
0.55–1.20	0.70
>1.20	0.75

$$d_0 + z_n - z_0 + \sum_{j=1}^{i} L_j \left(\frac{D_{sj}}{D_j}\right)^{5.26} \leq d_f$$

Optimization

Eliminating d_i between and dropping the constant terms yielded:

$$F = \sum_{i=1}^{n} \{k_m L_i D_i^m + [(n-i+1)k_h + k_{ei} L_i] L_i (\frac{D_{si}}{D_i})^{5.26}\}$$

$$+2\sum_{i=1}^{n-1}\left(\sum_{j=i+1}^{n} k_{ej}L_J\right)L_i\left(\frac{D_{si}}{D_i}\right)^{5.26}$$

If the constraint is loose, the cost function given by can be minimized. Assuming k_{ei} to be constant and differentiating with respect to D_i, equating it to zero, and simplifying gave the sewer diameters:

$$D_i = \{\frac{5.26D_{si}^{5.26}}{mk_m}[(n-i+1)k_h + k_{ei}L_i + 2\sum_{j=i=1}^{n} k_{ej}L_j]\}^{1/(m+5.26)},$$

$$D_n = \left[\frac{5.26D_{sn}^{5.26}}{mk_m}(k_n + k_{en}L_n)\right]^{1/(m+5.26)}$$

Using the diameters obtained by and, it was checked whether the assumption of, to be loose, was correct. If the check is satisfied, the diameters are in order. Otherwise, was made tight by assuming $d_n = d_f$. That is:

$$d_0 + z_n - z_0 + \sum_{i=1}^{n} L_i\left(\frac{D_{si}}{D_i}\right)^{5.26} - d_f = 0$$

Combining with by Lagrange multiplier λ, the merit function F1 was obtained:

$$F_1 = \sum_{i=1}^{n}\{k_m L_i D_i^m + [\lambda + (n-i+1)k_h + k_{ei}L_i]L_i(\frac{D_{si}}{D_i})^{5.26}\}$$

$$+2\sum_{i=1}^{n-1}\left(\sum_{j=i+1}^{n} k_{ej}L_j\right)L_i\left(\frac{D_{si}}{D_i}\right)^{5.26}$$

For optimality, differentiating F_1 in with respect to D_i, equating it to zero, and simplifying yielded:

$$D_i = \{\frac{5.26D_{si}^{5.26}}{mk_m}[\lambda + (n-i+1)k_h + k_{ei}L_i + 2\sum_{j=i+1}^{n} k_{ej}L_j]\}^{1/(m+5.26)} \quad ,i < n$$

$$D_n = [\frac{5.26D_{si}^{5.26}}{mk_m}(\lambda + k_h + k_{en}L_n]^{1/(m+5.26)}$$

In this case λ and D_i were obtained by solving and by trial and error. Thus, the sewer diameters were obtained. Using, the nodal depths di were obtained. The coefficients k_{ei} are functions of nodal depths. On the other hand, η_i and w_i depend on the sewer diameters. These variations were considered iteratively, and design diameters were reduced to the nearest available sizes.

Constraints Satisfaction

The optimization was carried out by considering the resistance equation as a tight constraint and the outfall depth constraint as a loose or tight constraint. However, some of the constraints, not considered so far, may be violated in the design so obtained. It is therefore essential to check the design against the constraint violation, and the violated constraint has to be made tight. The diameters thus obtained should satisfy the following constraints.

Depth Constraints

A sewer line should have a minimum cover depth h_{min}. The minimum cover depth is generally taken to be between 1 and 2 m. This is ensured by:

$$d_i \geq \max(d_{i-1,} d_i) + h_{\min}, \; 1,2,3....,n$$

Furthermore, a sewer line should not go very deep as the excavation at large depths involves dewatering problems. Thus, an upper limit d_{max} on the sewer depth was prescribed. That is:

$$d_i \geq d_{\max}$$

Velocity Constraints

Rounding off of the diameter alters η. Thus for obtaining the average velocity, it is necessary to find η. Using for the ith link, the altered η can be obtained. Designating this as ηi and using the continuity equation, the average velocity in the ith link V_i was obtained:

$$V_i \frac{4Q_i}{D_i^2[\cos^{-1}(1-2\eta_i)-2(1-2\eta_i)\sqrt{\eta_i(1-\eta_i)}]}$$

Swamee et al. suggested that the average velocity of flow exceeds the self-cleaning velocity for a specified time to flush the sediment deposits. On the other hand, Nalluri and Dabrowski gave a criterion for exceedence of the boundary shear stress. For simplicity, it is assumed that, to prevent clogging, a lower limit of the self-cleaning velocity Vsc is prescribed on V_i. That is:

$$V_i \geq V_{sc}$$

Table: Adopted Self-Cleaning Velocities.

Sewer diameter D (m)	Adopted self-cleaning velocity V_{sc} (m/s)
0.15–0.25	0.40
0.30–0.60	0.30
>0.60	0.20

Table: Maximum Recommended Velocities.

Sewer material	Scouring velocity Vsc (m/s)
Cast iron	3.5–4.5
Stoneware	3.0–4.5
Concrete	2.5–3.0
Brick	1.5–2.5

Similarly, an upper limit of the scouring velocity V_s is prescribed on the maximum achievable velocity occurring in the $_i$th link V_{imax} That is:

$$V_{i\max} \leq V_S$$

where $V_{i\max}$ was given by.

In the case of violation of the minimum cover constraint, the sewer line may be split at the violation node and designed as two sewer lines. This process forces the depth constraint at the splitting node. A similar method is applicable for the violation of the maximum depth constraint.

The self-cleaning velocity constraint is violated in the head reaches of the sewer line. The diameter of the violating link i can be obtained by considering it as a single link sewer line. Thus, using, the diameter of the constraint-violating link was obtained:

$$D_i = [\frac{5.26 D_{si}^{5.26}}{mk_m}(k_h + k_{en} L_i]^{1/(m+5.26)}$$

Similarly, using and, the nodal depth of d_i was found to be:

$$d_i = L_i \left(\frac{D_{si}}{D_i} \right)^{5.26} + d_{i-1} + z_i - z_{i-1}$$

where D_{si} was obtained by. As there is considerable reduction in the revised sewer diameter, the corresponding increase in the average velocity may satisfy the self-cleaning velocity constraint. If the constraint violation persists, keeping the same diameter, the slope of the link may be increased to achieve the self-cleaning velocity and the remaining part can be designed as a separate sewer line. Thus, using and, the revised slope was obtained:

$$S_{0i} = \frac{0.663 V_{sc}^2 \cos^{-1}(1-2\eta_i)}{[\cos^{-1}(1-2\eta_i) - (1-2\eta_i)\sqrt{\eta_i}(1-\eta_i)] g D_i}$$

$$.(1n\{\frac{\varepsilon \cos^{-1}(1-2\eta_i)}{3D_i[\cos^{-1}(1-2\eta_i) - 2(1-2\eta_i)\sqrt{\eta_i}(1-\eta_i)]}\})^{-2}$$

Using the revised slope and, the nodal depth *di* can be obtained. Alternatively, flushing arrangements may be made if the increased slope involves considerable excavation cost. Adopting a similar procedure for violation of the maximum velocity constraint, the revised slope was obtained:

$$S_{0i} = \frac{0.545V_s^2}{gD_i}[1n\left(\frac{\varepsilon}{3.65D_i}\right)]^{-2}$$

Link/Node Number I								
Entity	0	1	2	3	4	5	6	7
(a) Data								
Li (m)	—	75.0	80.0	65.0	90.0	85.0	90.0	95.0
zi (m)	100.00	101.00	100.75	100.15	100.00	100.50	100.25	99.50
qi (m3/s)	0.015	0.017	0.015	0.025	0.020	0.040	0.030	0.000
(b) Computations								
Qi (m3/s)	-	0.015	0.032	0.047	0.072	0.092	0.132	0.162
Di (m)a	-	0.3322	0.3935	0.4293	0.4706	0.4822	0.4992	0.4532
di (m)a	-	2.5176	2.3048	1.7460	1.6786	2.2906	2.2442	2.2821
Di (m)b	-	0.3260	0.3988	0.4361	0.4799	0.4950	0.5197	0.4990
di (m)b	-	2.5165	2.3012	1.7392	1.6637	2.2614	2.1760	2.0000
Di (m)c	-	0.350	0.400	0.450	0.500	0.500	0.500	0.500
di (m)c	-	2.520	2.300	1.740	1.660	2.260	2.180	2.000
η d	-	0.4960	0.6133	0.5503	0.5615	0.5749	0.6227	0.5895
Vimax (m/s)d	-	0.3595	0.4145	0.5716	0.6861	0.8439	1.0693	1.4279
Qimax (m/s)d	-	0.0323	0.0486	0.0849	0.1258	0.1547	0.1960	0.2617
Vi (m/s)d	-	0.3147	0.3960	0.5240	0.6343	0.7875	1.0269	1.3453
Invert level (m)	98.500	98.480	98.450	98.410	98.340	98.240	98.070	97.500

[a]Loose constraint, λ = 0.
[b]Tight constraint, λ = 142.360.
[c]Rounding off.
[d]Constraints checking and carrying capacity.

The new nodal depth di can be obtained by using and. A drop equal to the difference between the old and new di was provided at the node i. In this case, computation of cost using is not possible as there are two depths at the node i. This situation can be circumvented by introducing a fictitious link having zero length and the corresponding two depths at the end points. Thus, barring the corrected link and the fictitious link, the entire design remained the same.

Algorithm Description

The following steps describe the foregoing development:

- Initialize d_0 = 1.5 m, w_i = 1 m, n_i = 0.5, and k_h/c_e and $k_{ei} = 0.5c_e + 0.25c_r + c_s$.
- Using find k_{Di}, and using find D_{si}.
- Using and find D_i, and using Table find n_i.

- Using find S_{oi}, and using find d_i.
- Using and find w_i and revise $d_o = D_1 + 1$.
- Using find k_{ei}.
- Knowing the diameters, revise the average value of k_h/c_e.
- Repeat Steps 2–7 until two successive sewer profiles are close.
- If constraint is loose, go to the next step; otherwise, find λ by and and repeat Steps 3–9 until two successive λ are close.
- Reduce D_i to the nearest commercial size and round off d_i. Using find S_{oi}.
- Using find n, and using find V_i.
- Using find V_{imax}, and using find Q_{imax}.
- If any of the constraints are not satisfied, split the sewer line into two and design both the lines using Steps 1–12.
- For violation of, using and find the new nodal depth and provide the drop equal to the difference between the old and new nodal depths. Introduce a fictitious link of zero length.
- Find *F*.

Design Example

Design a cast iron sewer line for the data given in Table. The unit cost ratios are k_m/c_e = 10 SI units, c_r/c_e = 0.1 m^{-1}, c_s/c_e = 1 m, and k_h/c_e = 3 m^2. The depth at the outfall is 2 m.

The following values were assumed in the design process: m = 1.62; ε= 0.5 mm, and g = 9.8 m/s^2. Initializing w_i = 1 m and η_i = 0.5 for all the links and adopting d_o = 1.5 m, iterations as shown in Table were carried out.

Salient Points

It is possible to design a sewer network by repeated application of the numerical algorithm. Starting from the outfall point, moving opposite the flow direction, and following the maximum discharge sewer identify the primary sewer line in a sewerage network. On the other hand, starting from the junction points of the primary sewer line and following the path opposite to the flow direction identifies the secondary sewer lines. Similarly, the higher order sewer lines can be identified. These sewer lines can be designed separately by the numerical algorithm described herein. Therefore, the entire sewerage network can be decomposed to a set of sewer lines that can be designated as individual units. Thus, there is considerable saving in computational effort.

As in the majority of cases of the sewer line design, the constraint is loose and the diameters can be computed by straightforward application of and. Only in a few cases is the constraint tight and one has to resort to trial and error for obtaining the Lagrange multiplier λ.

Hydraulics of Sewers

When a fluid flows past a point or through a path different parameters associated with the flow of the fluid, certain parameters vary and others may remain constant. The two basic parameters of any fluid flow are velocity of the fluid particle or element and the pressure of the fluid at the point under consideration. The flow of fluids can be classified in different patterns based on the variation of the flow parameters with time and distance. The benefit of characterizing the fluid flow as certain patterns helps in analyzing it under the appropriate solution paradigm.

Classification based on Variation with Time

The classification of the fluid flow based on the variation of the fluid flow parameters with time characterizes the flow in two categories, steady and unsteady flow. If the flow parameters, such as velocity, pressure, density and discharge do not vary with time or are independent of time then the flow is steady. If the flow parameters vary with time then the flow is categorized as unsteady.

In real conditions it is very rare to have such flows with parameters exactly constant with time. The parameters usually vary with time but variation is within a small range such as the average of particular parameter is constant for certain duration of time.

Classification based on Variation with Space

The other classification criterion for the fluid flow is based on the variation of the flow parameters with distance or space. It characterizes the flow as uniform or non-uniform. The fluid flow is a uniform flow if the flow parameters remain constant with distance along the flow path and the fluid flow is non-uniform if the flow parameters vary and are different at different points on the flow path.

For a uniform flow, by its definition, the area of the cross section of the flow should remain constant. So a fitting example of the uniform flow is the flow of a liquid thorough a pipeline of constant diameter. And contrary to this the flow through a pipeline of variable diameter would be necessarily non-uniform.

Flow Types and Examples

A steady flow can be uniform or non-uniform and similarly an unsteady flow can also be uniform or non-uniform. For a steady flow discharge is constant with time and for a uniform flow the area of cross section of the fluid flow is constant through the flow path.

Examples of Different Flow Types

- Steady and Uniform Flow: Flow through a pipeline of constant diameter with a discharge constant with time.
- Steady and Non-Uniform Flow: Fixed discharge flow through a tapering pipe. Water flow

through a river with a constant discharge is also a good example of such flow as the span of river generally varies with distance and amount of water flow in river is constant.

- Unsteady and Uniform Flow: A flow through pipeline of constant cross section with sudden changes in fluid discharge or pressure.
- Unsteady and Non-Uniform Flow: Pressure surges in a flow through a pipe of variable cross section. A practical example can be the water flow in the network of canals during water release.

Formulae used in Hydraulic Design of Sewers

In principle, all open channel flow formulae can be used in hydraulic design of sewer pipes tough Manning's formula is the most common today.

- Chezy's Formula:

$$V = C\sqrt{R.S}$$

where V (m/s), R (m) and S (m/m)

Coefficient C given by Kutter reads as following in metric units:

$$C = \frac{(23 + 0.00155/s) + 1/n}{1 + \left(n/\sqrt{R}\right)(23 + 0.00155/s)}$$

n is same as Manning's equation.

or simplified form of this coefficient can be written as:

$$C = \frac{100\sqrt{R}}{m + \sqrt{R}}$$

m = 0.35 for concrete pipes

m = 0.25 for vitrified clay pipes

- Darcy-Weisbach Equation:

$$h_L = f\frac{L}{D}\frac{v^2}{2g}$$

- Manning's Equation:

$$V = \frac{1}{n}R^{2/3}S^{1/2}$$

N is same as the n in Kutter's formula. In general n = 0.013 - 0.015 for sewer pipes.

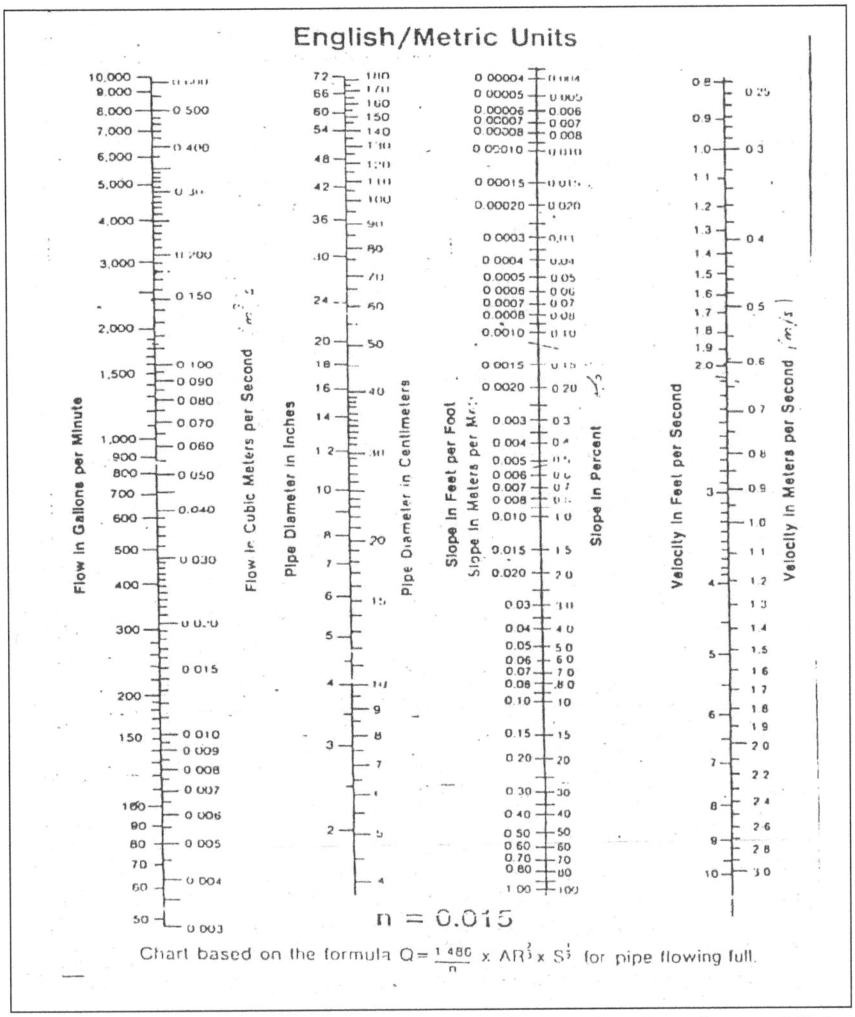

Figure: Nomogram for solution of Manning's equation for circular pipes flowing full (n = 0.015).

Hydraulics of Partially Filled Sections

Sanitary sewers are not to be designed to flow full. Thus "hydraulics of partially filled section" is important.

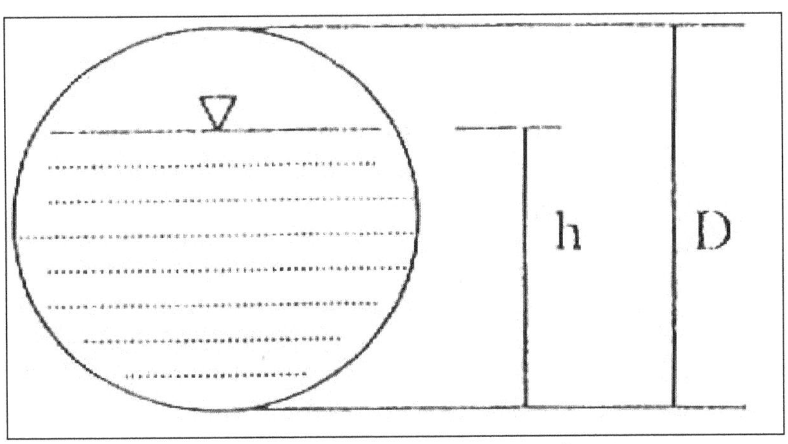

Figure: Partially filled section of a circular pipe.

Relationship between (Q, V, R, A), and (Qf , Vf , Rf , Af) are given in the form of diagram depending on h/D (d/D) ratio in the Figure where:

- Q, V, R, A: Flow rate, flow velocity, hydraulic radius and area for partly filled flow.
- Qf , Vf , Rf , Af: Flow rate, flow velocity, hydraulic radius and area for full flow condition.
- Remember that in the Figure, coefficient n varies with diameter and h/D ratio.

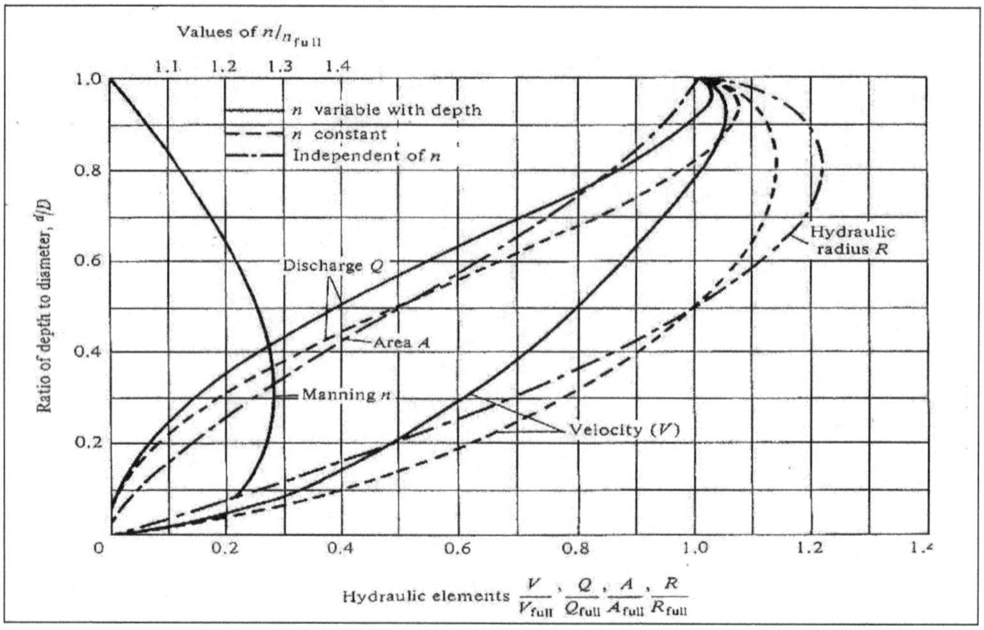

Figure: Hydraulic elements for circular pipes.

Hydraulic elements $\frac{V}{V_{full}}, \frac{Q}{Q_{full}}, \frac{A}{A_{full}}, \frac{R}{R_{full}}$

If the variation of coefficient n with depth is to be neglected, calculations involving flow in partly filled sewers can easily be handled using the data in Tables.

d^b	0.00	0.01	0.02	0.03	0.04	0.05	0.06	0.07	0.08	0.09
0.0	....	10.11	7.11	5.77	4.96	4.41	4.00	3.68	3.42	3.20
0.1	3.02	2.85	2.72	2.60	2.49	2.39	2.29	2.21	2.13	2.06
0.2	2.00	1.93	1.88	1.82	1.77	1.72	1.68	1.63	1.59	1.55
0.3	1.51	1.48	1.44	1.41	1.38	1.35	1.32	1.29	1.26	1.23
0.4	1.21	1.18	1.16	1.14	1.11	1.09	1.07	1.05	1.03	1.01
0.5	0.990	0.971	0.952	0.934	0.916	0.899	0.882	0.805	0.849	0.833
0.6	0.818	0.802	0.787	0.772	0.758	0.744	0.730	0.716	0.702	0.690
0.7	0.676	0.662	0.650	0.637	0.624	0.612	0.600	0.588	0.576	0.564
0.8	0.552	0.541	0.529	0.518	0.507	0.495	0.484	0.473	0.462	0.451
0.9	0.440	0.429	0.418	0.407	0.395	0.384	0.372	0.360	0.348	0.334
1.0	0.312									

Where,

- Q=flowrate, $m^{3/s}$.
- n= Manning coefficient of friction.
- d=depth of flow, m.
- S=slope of energy grade line, m/m.
- [b]D= diameter of conduit.m.

Table: Values of K' for circular channels in terms of depth of flow in the equation $Q = \left(\frac{K'}{n}\right) D^{8/3} S^{1/2}$

$\frac{d^b}{D}$	0.00	0.01	0.02	0.03	0.04	0.05	0.06	0.07	0.08	0.09
0.0	......	0.000047	0.00021	0.00050	0.00093	0.00150	0.00221	0.00306	0.00407	0.00521
0.1	0.00651	0.00795	0.00953	0.0113	0.0131	0.0152	0.0173	0.0196	0.0220	0.0246
0.2	0.0273	0.0301	0.0331	0.0362	0.0394	0.0427	0.0461	0.0497	0.0534	0.0572
0.3	0.0610	0.0650	0.0691	0.0733	0.0776	0.0820	0.0864	0.0910	0.0956	0.1003
0.4	0.1050	0.1099	0.1148	0.1197	0.1248	0.1298	0.1349	0.1401	0.1453	0.1506
0.5	0.156	0.161	0.166	0.172	0.177	0.183	0.188	0.193	0.0199	0.204
0.6	0.209	0.215	0.220	0.225	0.231	0.236	0.241	0.246	0.251	0.256
0.7	0.261	0.266	0.270	0.275	0.280	0.284	0.289	0.293	0.297	0.301
0.8	0.305	0.308	0.312	0.315	0.318	0.321	0.324	0.326	0.329	0.331
0.9	0.332	0.334	0.335	0.335	0.335	0.335	0.334	0.332	0.329	0.325
1.0	0.312									

Where,

- Q= flow rate, m^3/s.
- n=manning coefficient of friction.
- D=diameter of conduit.
- S=slope of energy grade line, m/m.
- [b]d=depth of flow.

Calculation of Hydraulic Elements by Trigonometry

The derivation of hydraulic elements can be done also by trigonometry and Figure will help to formulate the hydraulic elements formulae. In Figure, θ is wetted angle, h is water depth in the sewer and D is the diameter of the sewer.

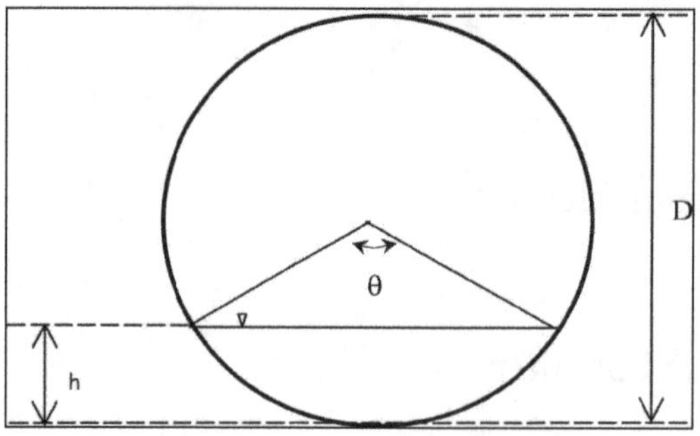

Figure: Hydraulic elements derivation.

(Q, V, R, A) and (Qf , Vf , Rf , Af) can be expressed in terms of wetted angle (θ).

$$\frac{h}{D} = \frac{1}{2}(1 - \cos\frac{\theta}{2}$$

$$\cos\frac{\theta}{2} = 1 - 2(\frac{h}{D}) = \varnothing$$

$$\theta = 2\cos^{-1}\varnothing$$

$$\frac{A}{A_{full}} = \frac{1}{2\pi}(\theta - \sin\theta) = \frac{1}{\pi}\cos^{-1}\varnothing - \frac{1}{\pi}\varnothing\sqrt{1-\varnothing^2}$$

$$\frac{R}{R_{full}} = \frac{\theta - \sin\theta}{\theta} = 1 - \frac{\varnothing\sqrt{1-\varnothing^2}}{\cos^{-1}\varnothing}$$

$$\frac{V}{V_{full}} = \left(\frac{R}{R_{full}}\right)^{2/3}$$

$$\frac{V}{V_{full}} = \left(\frac{\theta - \sin\theta}{\theta}\right)^{2/3}$$

$$\frac{Q}{Q_{full}} = \frac{1}{2\pi}\frac{(\theta - \sin\theta)^{5/3}}{\theta^{2/3}}$$

$$\frac{Q}{Q_{full}} = 0.159\theta^{-\frac{2}{3}}(\theta - \sin\theta)^{5/3}$$

Manhole

Manholes are masonry or RCC chambers constructed at suitable intervals along the sewer lines for providing access into them.

Purposes of Manhole

- They are used to carry out inspection, cleaning and removing obstruction in the sewer line.
- Manhole allows joining of sewers or changing the direction of sewer or alignment of sewer or both.

- They allow the escape of considerable gases through perforated cover and thus help in ventilation of sewage.
- They facilitate the laying of sewer line in convenient lengths.

Location of Manhole

Manhole is provided when:

- There is change in grade of sewer.
- There is change in alignment.
- There is change in size of sewer.
- At junction of two or more sewers.
- Manhole is also provided in straight alignment of sewers at regular intervals depending upon the diameters of sewers. It ranges from 90m to 150m (300' – 500') e.g. 75m for 60cmφ, 120m for 90cmφ and 150m for 120cmφ.

Types of Manhole

Depending upon the depth the manhole can be classified as:

- Shallow Manhole: It is provided at shallow depth of 75-90cm (2'-3'). It is provided at the beginning of branch sewer or at a place not subjected to heavy traffic. It is provided with a light cover at its top it is also called inspection chamber.
- Normal Manhole: It is provided in sewer line at depth of 150cm with a heavy cover on its top. It is generally of square shape (or rectangular shape).
- Deep Manhole: They are provided at depth greater than 150cm with heavy cover at the top. The size is gradually increased and a facility for going down is provided.

Packaged Metering Manholes

The Packaged Metering Manhole was introduced to the wastewater market. in 1982. In its simplest form it consists of a flume molded into a composite (FRP) manhole. The full range of flumes may be integrally molded into the composite manhole, along with inlet and outlet adapters, pipe stubs, and monitoring accessories for a complete package. Packaged Metering Manholes (PMMs) provide a complete and water tight flow monitoring package that can easily be installed during initial construction or as a retrofit when there is a need to measure flow in an underground sewer pipe without access to an existing manhole.

For installation on existing sewer lines an area at least the size of the manhole is excavated and the existing sewer line is cut to match the length of the pipe stubs on the manhole. A concrete pad is then poured or set in place to provide a level surface for the PMM and to act as an anchoring device to overcome buoyant forces acting on the manhole. The PMM is set on the pad, leveled with shims and anchored in place. The typical method of connecting the existing sewer pipe to the pipe stubs on the PMM is with neoprene boots who offer some flexibility and forgiveness in lining up the pipe to the pipe stubs. The area around the manhole is then backfilled and you are left with easily accessible flow monitoring package.

PMMs interiors are finished with a premium grade marine white gel coat that provides a smooth non-porous surface to resist bacteria for a cleaner environment, and to maximize illumination for a safer workspace.

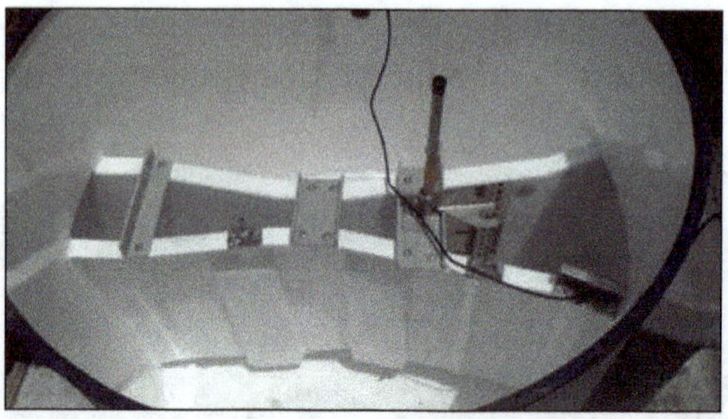

During installation, the void space on the sides of the flume are typically filled with sand and capped with grout with a slope back towards the flume. This allows all water to flow back into the flume and out the manhole in the event of a sewer backup.

PMMs are built to meet the requirements of ANSI/ASTM D-3753 "Standard Specifications for Glass-Fiber-Reinforced Polyester Manholes and Wetwells." Packaged Metering Manholes are available with:

- All flume types and sizes.

- Hinged lid or concentric manhole reducer with H-20 rated lids or non-traffic load manhole covers.
- FRP Ladder standard on 4' and deeper manholes (exceeds OSHA Safety Standard 1910.27).
- All standard flume accessories (i.e. sample pipe, bubble pipe, ultrasonic flow meter bracket, stilling wells, pH probe lift out brackets, etc.).
- Safety rails, harness, and access poles.
- Air vents.

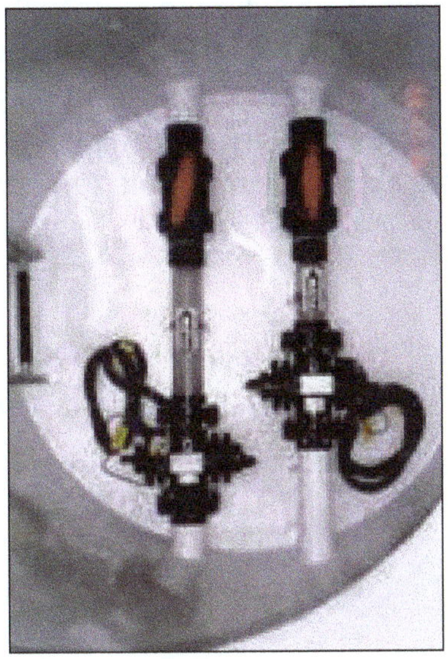

In addition to flumes, Packaged Metering Manholes can be equipped with pipe connections for installing magnetic flow meters or other pressurized line type flow meters.

Domed-Top Manholes

Domed-top manholes provide a wide range of benefits, including durability and ease of access. That being said, installing a domed-top manhole isn't always the right choice, meaning it's a good idea to examine this issue in a little closer detail.

When you Shouldn't use a Domed-Top Manhole

Basically, if you're installing a manhole in an area where traffic is likely, whether pedestrian or vehicular, you should not use a domed top manhole. The reason for this is very simple: Domed-top manholes will protrude a foot above grade. Because domed-top manholes are elevated above grade, they should only be used in non-traffic areas. Installing this style of manhole in a high-traffic area while exposing your manhole to damage and may threaten the safety of passersby.

Domed-Top Manhole Benefits

Now that you now that you can only use domed-top manholes in an area with no traffic, it's time to take a look at some of the biggest benefits of choosing this style of manhole.

First and foremost, because the domed-top manhole rises above the ground, it helps prevent water from infiltrating your manhole, meaning that your equipment will be protected from moisture damage. Another big benefit of domed-top manholes is that they provide easy entry and exit for your operators, meaning they can quickly take readings or perform necessary maintenance. As long as you install your domed-top manhole in the right locations, you should only benefit from this versatile tool.

How the Dome is Attached

If you're interested in installing a domed top manhole in your system, it's a good idea to learn a little bit about how the domed-top will be attached to your manhole. To make sure the domed-top can easily open and close while staying secure; it will be attached to your manhole using a piano hinge made from stainless steel. Generally, this hinge will be connected to a hinge block that is mounted in your manhole. To improve security, many people choose to add a padlock to their domed-top. A gas strut can be used to keep the manhole top open.

References

- Sewage-management-and-its-benefits-to-man-319523571: researchgate.net, Retrieved 25, August 2020
- Sewers-and-Types-of-Sewers, civil-engineering: idc-online.com, Retrieved 10, March 2020
- Design-of-Sewer-Line-245300325: researchgate.net, Retrieved 16, June 2020

Chapter 5

Sanitation of Sewage

Sewage is the waste matter or liquid refuse carried off by sewers. They pose serious health and environmental risks if not handled and properly disposed. This chapter takes a closer look at sanitation processes like primary, secondary and tertiary treatments, the different sewage systems and sewage handling methods in order to provide a clear understanding of the subject matter.

Sewage Sanitation

For safe disposal of the sewage generated from a locality efficient collection, conveyance, adequate treatment and proper disposal of treated sewage is necessary. To achieve this, following conditions should be satisfied:

1. Sewage should not pollute the drinking water source, either surface or groundwater, or water bodies that are used for bathing or recreational purposes.
2. The untreated sewage during conveyance should not be exposed so as to have access to human being or animals and should not give unsightly appearances or odour nuisance, and should not become a place for breeding flies.
3. It should not cause harm to public health and adversely affect the receiving environment.

The collection system is meant for collection of the sewage generated from individual houses and transporting it to a common point where it can be treated as per the needs before disposal. In olden days, waste generated from water closets was collected by conservancy methods and other liquid waste was transported through open drain to finally join natural drains. Since, the excreta was carried through carts, it was not hygienic method for transportation to the disposal point. Now, collection and conveyance of sewage is done in water carriage system, where it is transported in closed conduit using water as a medium.

Types of Sanitary Sewer Systems

Combined Sewer

A combined sewer is a sewage collection system of pipes and tunnels designed to also collect surface runoff. Combined sewers can cause serious water pollution problems during combined sewer overflow (CSO) events when wet weather flows exceed the sewage treatment plant capacity. This type of sewer design is no longer used in building new communities (because current design separates sanitary sewers from runoff), but many older cities continue to operate combined sewers.

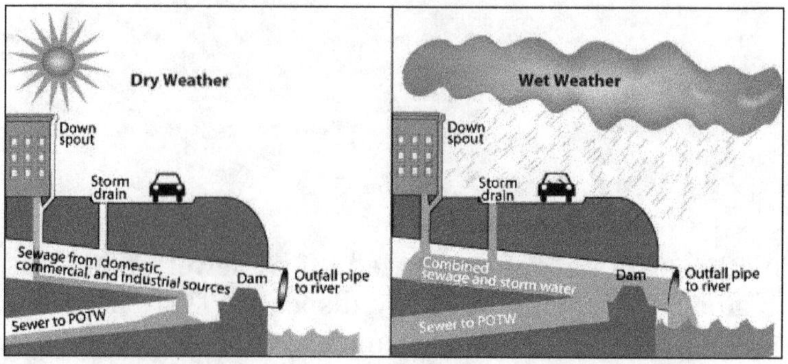

Combined Sewer System. During dry weather (and small storms), all flows are handled by the publicly owned treatment works (POTW). During large storms, the relief structure allows some of the combined stormwater and sewage to be discharged untreated to an adjacent water body.

Background

Recent archaeological discoveries have shown that some of the earliest sewer systems were developed 2500 BC in the ancient city of Harappa. The primitive sewers were carved in the ground alongside buildings. This discovery reveals the conceptual understanding of waste disposal by the early civilizations.

The earliest sewers were designed to carry street runoff away from inhabited areas and into surface waterways without treatment. Open sewers, consisting of gutters and urban streambeds, were common worldwide before the 20th century. In the majority of developed countries, large efforts were made during the late 19th and early 20th centuries to cover the formerly open sewers, converting them to closed systems with cast iron, steel, or concrete pipes, masonry, and concrete arches. Most sewage collection systems of the 19th and early to mid 20th century used single-pipe systems that collect both sewage and urban runoff from streets and roofs. This type of collection system is referred to as a combined sewer system. The rationale for combining the two was that it would be cheaper to build just a single system. Most cities at that time did not have sewage treatment plants, so there was no perceived public health advantage in constructing a separate "surface water sewerage" (the term used in the UK) or "storm sewer" (the term used in the U.S.) system.

When constructed, combined sewer systems were typically sized to carry three to 160 times the average dry weather sewage flows. It is generally infeasible to treat the volume of mixed sewage and surface runoff flowing in a combined sewer during peak runoff events caused by snowmelt or convective precipitation. As cities built sewage treatment plants, those plants were typically built to treat only the volume of sewage flowing during dry weather. Relief structures were installed in the collection system to bypass untreated sewage mixed with surface runoff during wet weather, protecting sewage treatment plants from damage caused if peak flows reached the headworks.

Combined Sewer Overflows (CSOs)

These relief structures, called storm-water regulators (in American English - or combined sewer overflows in British English) are constructed in combined sewer systems to divert flows in excess of the peak design flow of the sewage treatment plant. Combined sewers are built with control sections establishing stage-discharge or pressure differential-discharge relationships which may be either predicted or calibrated to divert flows in excess of sewage treatment plant capacity. A leaping weir may be used as a regulating device allowing typical dry-weather sewage flow rates to fall into an interceptor sewer to the sewage treatment plant, but causing a major portion of higher flow rates to leap over the interceptor into the diversion outfall. Alternatively, an orifice may be sized to accept the sewage treatment plant design capacity and cause excess flow to accumulate above the orifice until it overtops a side-overflow weir to the diversion outfall.

Photo of the interior of a combined sewer in Brighton, England.

CSO statistics may be confusing because the term may describe either the number of events or the number of relief structure locations at which such events may occur. A CSO event, as the term is used in American English, occurs when mixed sewage and stormwater are bypassed from a combined sewer system control section into a river, stream, lake, or ocean through a designed diversion outfall, but without treatment. Overflow frequency and duration varies both from system to system, and from outfall to outfall, within a single combined sewer system. Some CSO outfalls discharge infrequently, while others activate every time it rains.

CSOs should not be confused with sanitary sewer overflows. Sanitary sewer overflows are caused by sewer system obstructions, damage, or flows in excess of sewer capacity (rather than treatment plant capacity.) Sanitary sewer overflows may occur at any low spot in the sewer system rather than at the CSO relief structures. Absence of a diversion outfall often causes sanitary sewer overflows to flood residential structures and/or flow over traveled road surfaces before reaching natural drainage channels. Sanitary sewer overflows may cause greater health risks and environmental damage than CSOs if they occur during dry weather when there is no precipitation runoff to dilute and flush away sewage pollutants.

The storm water component contributes pollutants to CSO; but a major faction of pollution is the first foul flush of accumulated biofilm and sanitary solids scoured from the dry weather wetted perimeter of combined sewers during peak flow turbulence. Each storm is different in the quantity and type of pollutants it contributes. For example, storms that occur in late summer, when it has not rained for a while, have the most pollutants. Pollutants like oil, grease, fecal coliform from pet and wildlife waste, and pesticides get flushed into the sewer system. In cold weather areas, pollutants from cars, people and animals also accumulate on hard surfaces and grass during the winter and then are flushed into the sewer systems during heavy spring rains.

CSOs in the United States

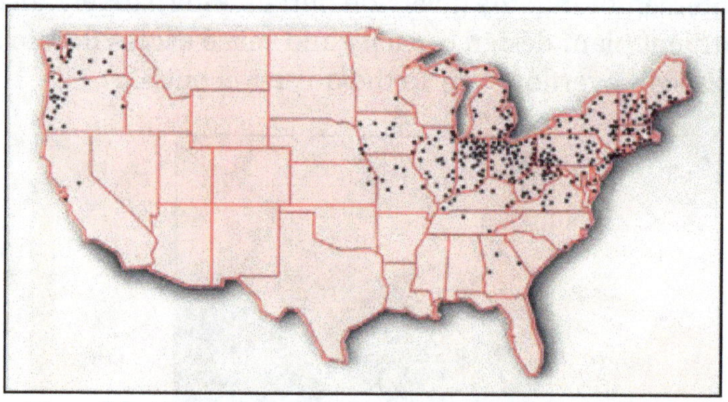

Most of the US combined sewer systems are in the Northeast and Great Lakes regions, and the Pacific Northwest.

About 860 communities in the US have combined sewer systems, serving about 40 million people. CSO discharges during heavy storms can cause serious water pollution problems in these communities. Pollutants from CSO discharges can include bacteria and other pathogens, toxic chemicals, and debris. These pollutants have also been linked with antimicrobial resistance, posing serious public health concerns. The U.S. Environmental Protection Agency (EPA) issued a policy in 1994 requiring municipalities to make improvements to reduce or eliminate CSO-related pollution problems. It is managed by the National Pollutant Discharge Elimination System (NPDES) permit program. The policy defined water quality parameters for the safety of an ecosystem; it allowed for action that are site specific to control CSOs in most practical way for community; it made sure the CSO control is not beyond a community's budget; and allowed water quality parameters to be flexible, based upon the site specific conditions. The CSO Control Policy required all states to have "nine minimum controls" in place by January 1, 1997, to decrease the effects of sewage overflow by making small improvements in existing processes. In 2000 Congress amended the Clean Water Act to require the municipalities to comply with the EPA policy.

Mitigation of CSOs

The United Kingdom Environment Agency identified unsatisfactory intermittent discharges and issued an Urban Wastewater Treatment Directive requiring action to limit pollution from combined sewer overflows. In 2009 the Canadian Council of Ministers of the Environment adopted a Canada-wide Strategy for the Management of Municipal Wastewater Effluent including national standards to (1) remove floating material from combined sewer overflows, (2) prevent combined sewer overflows during dry weather, and (3) prevent development or redevelopment from increasing frequency of combined sewer overflows.

Municipalities in the US have been undertaking projects to mitigate CSO since the 1990s. For example, prior to 1990, the quantity of untreated combined sewage discharged annually to lakes, rivers and streams in southeast Michigan was estimated at more than 30 billion US gallons (110,000,000 m^3) per year. In 2005, with nearly $1 billion of a planned $2.4 billion CSO investment put into operation, untreated discharges have been reduced by more than 20 billion US gallons (76,000,000 m^3) per year. This investment that has yielded an 85 percent reduction in CSO has included numerous sewer separation, CSO storage and treatment facilities and wastewater treatment plant improvements constructed by local and regional governments.

Many other areas in the US are undertaking similar projects. Cities like Pittsburgh, Seattle, Philadelphia, and New York are focusing on these projects partly because they are under federal consent decrees to solve their CSO issues. Both up-front penalties and stipulated penalties are utilized by EPA and state agencies to enforce CSO-mitigating initiatives and the efficiency of their schedules. Municipalities' sewage departments, engineering and design firms, and environmental organizations offer different approaches to potential solutions.

Sewer Separation

Some US cities have undertaken sewer separation projects — building a second piping system for all or part of the community. In many of these projects, cities have been able to separate only portions of their combined systems. High costs or physical limitations may preclude building a completely separate system. In 2011 Washington, D.C. separated its sewers in four small neighborhoods at a cost of $11 million. (The project cost also includes improvements to the drinking water piping system.)

CSO Storage

Another solution is to build a CSO storage facility, such as a tunnel that can store flow from many sewer connections. Because a tunnel can share capacity among several outfalls, it can reduce the total volume of storage that must be provided for a specific number of outfalls. Storage tunnels store combined sewage but do not treat it. When the storm is over, the flows are pumped out of the tunnel and sent to a wastewater treatment plant. One of the main concerns with CSO storage is the time it is stored before it is released. Without careful management of this time the water in the CSO storage facility runs the risk of going septic.

Washington, D.C. is building underground storage capacity as its primary strategy to address CSOs. As of 2011, the city is building a deep storage tunnel, adjacent to the Anacostia River, that will reduce overflows to the river by 98 percent, and 96 percent system-wide. (The city's overall "Clean Rivers" project, projected to cost $2.6 billion, includes other components, such as reducing stormwater flows.) The South Boston CSO Storage Tunnel is a similar project, completed in 2011.

Expanding Sewage Treatment Capacity

Some cities have expanded their basic sewage treatment capacity to handle some or all of the CSO volume. In 2002 litigation forced the city of Toledo, Ohio to double its treatment capacity and build a storage basin in order to eliminate most overflows. The city also agreed to study ways to reduce stormwater flows into the sewer system.

Retention Basins

Retention treatment basins or large concrete tanks that store and treat combined sewage are another solution. These underground structures can range in storage and treatment capacity from

2 million US gallons (7,600 m³) to 120 million US gallons (450,000 m³) of combined sewage. While each facility is unique, a typical facility operation is as follows. Flows from the overloaded sewers are pumped into a basin that is divided into compartments. The first flush compartment captures and stores flows with the highest level of pollutants from the first part of a storm. These pollutants include motor oil, sediment, road salt, and lawn chemicals (pesticides and fertilizers) that are picked up by the stormwater as it runs off roads and lawns. The flows from this compartment are stored and sent to the wastewater treatment plant when there is capacity in the interceptor sewer after the storm. The second compartment is a treatment or flow-through compartment. The flows are disinfected by injecting sodium hypochlorite, or bleach, as they enter this compartment. It then takes about 20-30 minutes for the flows to move to the end of the compartment. During this time, bacteria are killed and large solid materials settle out. At the end of the compartment, any remaining sanitary trash is skimmed off the top and the treated flows are discharged into the river or lake.

Screening and Disinfection Facilities

Screening and disinfection facilities treat CSO without ever storing it. Called "flow-through" facilities, they use fine screens to remove solids and sanitary trash from the combined sewage. Flows are injected with sodium hypochlorite for disinfection and mixed as they travel through a series of fine screens to remove debris. The fine screens have openings that range in size from 4 to 6 mm, or a little less than a quarter inch. The flow is sent through the facility at a rate that provides enough time for the sodium hypochlorite to kill bacteria. All of the materials removed by the screens are then sent to the sewage treatment plant through the interceptor sewer.

Reducing Stormwater Flows

Communities may implement low impact development techniques to reduce flows of stormwater into the collection system. This includes:

- constructing new and renovated streets, parking lots and sidewalks with interlocking stones, permeable paving and pervious concrete.
- installing green roofs on buildings.
- installing bioretention systems, also called rain gardens, in landscaped areas.
- Rainwater harvesting equipment collects runoff from building roofs during wet weather for irrigating landscapes and gardens during dry weather.
- Graywater collection and use on site reduces sewage discharges at all times.

Gray vs. Green Infrastructure

CSO mitigating initiatives that are solely composed of sewer system reconstruction are referred to as gray infrastructure, while techniques like permeable pavement and rainwater harvesting are referred to as green infrastructure. Conflict often occurs between a municipality's sewage authority and its environmentally active organizations between gray and green infrastructural plans.

The 2004 EPA *Report to Congress* on CSO's provides a review of available technologies to mitigate CSO impacts.

New Approaches

Smart Infrastructure

Recent technological advances in sensing and control have enabled the implementation of Real Time Decision Support Systems (RT-DSS) for CSO mitigation. Through the use of internet of things technology and cloud computing, CSO events can now be mitigated by dynamically adjusting setpoints for movable gates, pump stations, and other actuated assets in sewers and storm water management systems. Similar technology, called adaptive traffic control is used to control the flow of vehicles through traffic lights. RT-DSS systems take advantage of storm temporal and spatial variability as well as varying concentration times due to diverse land uses across the sewershed to coordinate and optimize control assets. By maximizing storage and conveyance RT-DSS are able to minimize overflows using existing infrastructure. Successful implementations of RT-DSS have been carried out throughout the United States and Europe.

History

As a product of the Industrial Revolution, many cities in Europe and North America grew in the 19th century, frequently leading to crowding and increasing concerns about public health. As part of a trend of municipal sanitation programs in the late 19th and 20th centuries, many cities constructed extensive sewer systems to help control outbreaks of disease such as typhoid and cholera. Initially these systems discharged sewage directly to surface waters without treatment. As pollution of water bodies became a concern, cities added sewage treatment plants to their systems. Most cities in the Western world added more expensive systems for sewage treatment in the early 20th century.

As Britain was the first country to industrialize, it was also the first to experience the disastrous consequences of major urbanisation and was the first to construct a sewerage system as we know it today to mitigate the resultant unsanitary conditions. Joseph Bazalgette designed an extensive underground sewerage system that diverted waste to the Thames Estuary, downstream of the main centre of population. Six main interceptor sewers, totalling almost 135 miles (217 km) in length, were constructed. The intercepting sewers, constructed between 1859 and 1865, were fed by 450 miles (720 km) of main sewers that, in turn, conveyed the contents of some 13,000 miles (21,000 km) of smaller local sewers. With only minor modifications, Bazalgette's engineering achievement remains the basis for sewerage design up into the present day.

In France, the Paris cholera epidemic of 1832 sharpened the public awareness of the necessity for some sort of drainage system to deal with sewage and waster water in a better and healthier way since the Seine received up to 100,000 cubic meters of wastewater per day. Between 1865 and 1920 Eugene Belgrand led the development of a large scale system for water supply and wastewater management. By 1894 laws were passed which made drainage mandatory. The treatment of Paris sewage, though, was left to natural devices as 5,000 hectares of land were used to spread the waste out to be naturally purified.

Society and Culture

A combined sewer-pipe being laid by the city's sewerage company in Ghent, Belgium.

The image of the sewer recurs in European culture as they were often used as hiding places or routes of escape by the scorned or the hunted, including partisans and resistance fighters in World War II. Fighting erupted in the sewers during the Battle of Stalingrad. The only survivors from the Warsaw Uprising and Warsaw Ghetto made their final escape through city sewers. Some have commented that the engravings of imaginary prisons by Piranesi were inspired by the Cloaca Maxima, one of the world's earliest sewers.

Regulations

United Kingdom

There is in the UK a legal difference between a storm sewer and a surface water sewer. You do not have a right of connection to a storm sewer under section 106 of the water Industry Act. These are normally the pipe line that discharges to a watercourse, downstream of a combined sewer overflow. It takes the excess flow from a combined sewer. A surface water sewer conveys rainwater; legally you have a right of connection for your rainwater to this public sewer. A public storm water sewer can discharge to a public surface water, but not the other way around, without a legal change in sewer status by the water company.

In Fiction

The theme of traveling through, hiding, or even residing in combined sewers is a common plot device in media. Famous examples of sewer dwelling are the Teenage Mutant Ninja Turtles, Stephen King's *It*, *Les Miserables*, *The Third Man*, *Ladyhawke*,*Mimic*, *The Phantom of the Opera*, *Beauty and the Beast*, and *Jet Set Radio Future*. The Todd Strasser novel *Y2K-9: the Dog Who Saved the World* is centered around a dog thwarting terroristic threats to electronically sabotage American sewage treatment plants.

Sewer Alligators

A well-known urban legend, the sewer alligator, is that of giant alligators or crocodiles residing in combined sewers, especially of major metropolitan areas. Two public sculptures in New York depict an alligator dragging a hapless victim into a manhole.

Alligators have been known to get into combined storm sewers in the southeastern United States. Closed-circuit television by a sewer repair company captured an alligator in a combined storm sewer on tape.

Sanitary Sewer

PVC sanitary sewer installation. Sanitary sewers are sized to carry the amount of sewage generated by the collection area. Sanitary sewers are much smaller than combined sewers designed to also carry surface runoff.

A sanitary sewer or "foul sewer" is an underground carriage system specifically for transporting sewage from houses and commercial buildings through pipes to treatment or disposal. Sanitary sewers are part of an overall system called sewerage or sewage system.

Sewage may be treated to control water pollution before discharge to surface waters. Sanitary sewers serving industrial areas also carry industrial wastewater.

Separate sanitary sewer systems are designed to transport sewage alone. In municipalities served by sanitary sewers, separate storm drains may be constructed to convey surface runoff directly to surface waters. Sanitary sewers are distinguished from combined sewers, which combine sewage with stormwater runoff in the same pipe. Sanitary sewer systems are considered beneficial because they avoid combined sewer overflows.

Background

Sewage treatment is less effective when sanitary waste is diluted with stormwater, and combined sewer overflows occur when runoff from heavy rainfall or snowmelt exceeds hydraulic capacity of sewage treatment plants. To overcome these disadvantages, some cities built separate sanitary sewers to collect only municipal wastewater and exclude stormwater runoff collected in separate storm drains. The decision between a combined sewer system or two separate systems is mainly based on need for sewage treatment and cost of providing treatment during heavy rain events. Many cities with combined sewer systems built prior to sewage treatment have not replaced those sewer systems.

Types

Conventional Gravity Sewers

In the developed world, sewers are pipes from buildings to one or more levels of larger underground trunk mains, which transport the sewage to sewage treatment facilities. Vertical pipes, usually made of precast concrete, called manholes, connect the mains to the surface. Depending upon site application and use, these vertical pipes can be cylindrical, eccentric or concentric. The manholes are used for access to the sewer pipes for inspection and maintenance, and as a means to vent sewer gases. They also facilitate vertical and horizontal angles in otherwise straight pipelines.

Manhole access to sewer; person shows scale.

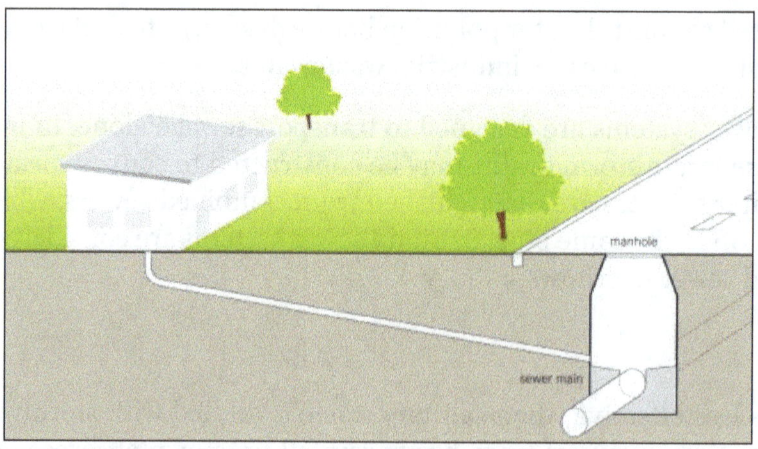

Schematic of a conventional sanitary sewer to convey blackwater and greywater from households to a centralized sewage treatment facility.

Pipes conveying sewage from an individual building to a common gravity sewer line are called laterals. Branch sewers typically run under streets receiving laterals from buildings along that street and discharge by gravity into trunk sewers at manholes. Larger cities may have sewers called interceptors receiving flow from multiple trunk sewers.

A manhole cover for a sanitary sewer access point.

View looking down into an open manhole showing two converging sanitary sewer lines. The larger line enters from the right and changes direction within the manhole to exit from the top of the photo. A smaller line enters from the bottom of the photo under the access steps. The concrete floor of the manhole has channels to minimize accumulation of solids.

Interior photo of a large sanitary sewer from an access manhole.

Design and sizing of sanitary sewers considers the population to be served over the anticipated life of the sewer, per capita wastewater production, and flow peaking from timing of daily routines. Minimum sewer diameters are often specified to prevent blockage by solid materials flushed down toilets; and gradients may be selected to maintain flow velocities generating sufficient turbulence to minimize solids deposition within the sewer. Commercial and industrial flows are also considered, but diversion of surface runoff to storm drains eliminates wet weather flow peaks of inefficient combined sewers.

Force mains

Pumps may be necessary where gravity sewers serve areas at lower elevations than the sewage treatment plant, or distant areas at similar elevations. A lift station is a gravity sewer sump with a pump to lift accumulated sewage to a higher elevation. The pump may discharge to another gravity sewer at that location or may discharge through a pressurized force main to some distant location.

Effluent sewer

Effluent sewer systems, also called septic tank effluent drainage (STED) or solids-free sewer (SFS) systems, have septic tanks that collect sewage from residences and businesses, and the effluent that comes out of the tank is sent to either a centralized sewage treatment plant or a distributed treatment system for further treatment. Most of the solids are removed by the septic tanks, so the treatment plant can be much smaller than a typical plant. In addition, because of the vast reduction in solid waste, a pumping system can be used to move the wastewater rather than a gravity system. The pipes have small diameters (typically 1.5" to 4" (inches), or 4 to 10 cm). Because the wastestream is pressurized, they can be laid just below the ground surface along the land's contour.

Simplified Sewer

Simplified sanitary sewers consist of small-diameter pipes (typically 100 mm or about 4 inches), often laid at fairly flat gradients (1 in 200). The investment cost for simplified sanitary sewers can be about half the cost of conventional sewers. However, the requirements for operation and maintenance are usually higher. Simplified sewers are most common in Brazil and are also used in a number of other developing countries.

Vacuum Sewer

In low-lying communities, wastewater may be conveyed by vacuum sewer. Pipelines range in size from pipes of six inches (150 mm) in diameter to concrete-lined tunnels of up to thirty feet (10 m) in diameter. A low pressure system uses a small grinder pump located at each point of connection, typically a house or business. Vacuum sewer systems use differential atmospheric pressure to move the liquid to a central vacuum station.

Maintenance

Sanitary sewer overflow can occur due to blocked or broken sewer lines, infiltration of excessive stormwater or malfunction of pumps. In these cases untreated sewage is discharged from a sanitary sewer into the environment prior to reaching sewage treatment facilities. To avoid this, maintenance is required.

The maintenance requirements vary with the type of sanitary sewer. In general, all sewers deteriorate with age, but infiltration/inflow is a problem unique to sanitary sewers, since both combined sewers and storm drains are sized to carry these contributions. Holding infiltration to acceptable levels requires a higher standard of maintenance than necessary for structural integrity considerations of combined sewers. A comprehensive construction inspection program is required to prevent inappropriate connection of cellar, yard, and roof drains to sanitary sewers.

The probability of inappropriate connections is higher where combined sewers and sanitary sewers are found in close proximity, because construction personnel may not recognize the difference. Many older cities still use combined sewers while adjacent suburbs were built with separate sanitary sewers.

For decades, when sanitary sewer pipes cracked or experienced other damage, the only option was an expensive excavation, removal and replacement of the damaged pipe, typically requiring street repavement afterwards. In the mid-1950s a unit was invented where two units at each end with a special cement mixture in between was pulled from one manhole cover to the next, coating the pipe with the cement under high pressure which then cured rapidly, sealing all cracks and breaks in the pipe. Today, a similar method using epoxy resin is used by some municipalities to re-line aging or damaged pipes, effectively creating a "pipe in a pipe". These methods may be unsuitable for locations where the full diameter of the original pipe is required to carry expected flows, and may be an unwise investment if greater wastewater flows may be anticipated from population growth, increased water use, or new service connections within the expected service life of the repair.

Another popular method for replacing aged or damaged lines is called Pipe bursting, where a new pipe (typically PVC or ABS) is drawn thru the old pipe behind an "expander head" that breaks apart the old pipe as the new one is drawn through behind it.

These methods are most suitable for trunk sewers, since repair of lines with lateral connections is complicated by making provisions to receive lateral flows without accepting undesirable infiltration from inadequately sealed junctions.

History

Sanitary sewers evolved from combined sewers built where water was plentiful. Animal feces accumulated on city streets while animal-powered transport moved people and goods. Accumulations of animal feces encouraged dumping chamber pots into streets where night soil collection was impractical. Combined sewers were built to use surface runoff to flush waste off streets and move it underground to places distant from populated areas. Sewage treatment became necessary as population expanded, but treatment of diluted waste from combined sewers is more expensive than treating undiluted sewage.

Simplified Sewerage

Simplified sewerage, also called small-bore sewerage, is a sewer system that collects all household wastewater (blackwater and greywater) in small-diameter pipes laid at fairly flat gradients. Simplified sewers are laid in the front yard or under the pavement (sidewalk) or - if feasible - inside the back yard, rather than in the centre of the road as with conventional sewerage. It is suitable for existing unplanned low-income areas, as well as new housing estates with a regular layout. It allows for a more flexible design. With simplified sewerage it is crucial to have management arrangements in place to remove blockages, which are more frequent than with conventional sewers. It has been estimated that simplified sewerage reduces investment costs by *up to* 50% compared to conventional sewerage.

Simplified sewerage is sometimes also referred to as conventional sewerage with appropriate standards, implying that most conventional sewers are overdesigned.

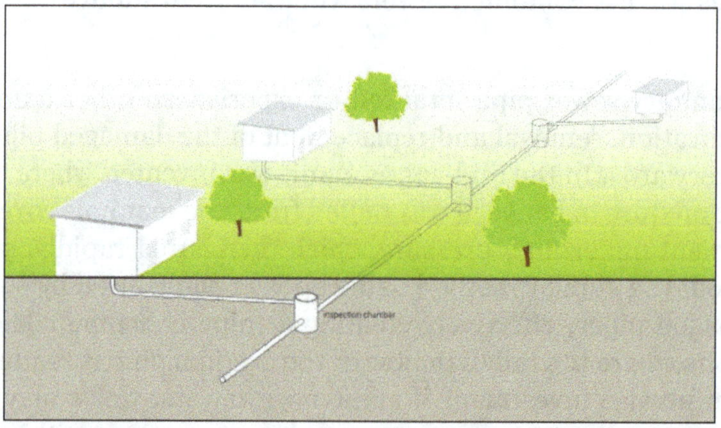

Schematic of a simplified sewer: Smaller diameter pipes are laid at a shallower depth and at a flatter gradient than for conventional sewers.

The concept of simplified sewerage emerged in parallel in Natal, Brazil and Karachi, Pakistan in the early 1980s without any interaction or communication. In both cases particular emphasis was given to community mobilization, an essential element for the success of simplified sewerage. In Latin America, and particularly in Brazil, simplified sewerage is also known as condominial sewerage, a term that undercores the importance of community participation in planning and maintenance at the level of a housing block (known as *condominio* in the Spanish and Portuguese use of the term).

Background

In developing countries, connection to sewer systems is often costly for poor households, despite typically low monthly sewer tariffs. This apparent paradox is explained by the high costs of in-plot and in-house sanitary installations that have to be paid entirely by the user, by sometimes high sewer connection fees levied by utilities, and by a lack of community consultation. As a result, in many cities in developing countries conventional sewers are laid at high costs under a street, while many users on that street do not connect to them. In Brazil, in some cities connection rates in the early 1990s were less than 40% of the intended beneficiary population.

Application

Simplified sewerage is most widely used in Brazil. It is estimated that in Brazil some 5 million people in over 200 towns and cities are served with simplified sewerage - or condominial sewerage. This corresponds to about 3% of the population of Brazil and about 6% of the population connected to sewers. They serve poor and rich alike.

Simplified sewerage has also been used in

- Bolivia, beginning with a pilot project in El Alto;
- Honduras, primarily in marginal areas of Tegucigalpa where simplified sewerage has been introduced in 20 communities with 24,000 inhabitants;

- Peru, primarily in marginal areas of Lima;
- in South Africa, where pilot projects were carried out in Johannesburg and Durban;
- in Sri Lanka, where the National Housing Development Authority implemented over 20 schemes in the 1980s and 90s.

In Pakistan, beginning with the Orangi Pilot Project in Karachi, a variation of simplified sewerage using larger diameter pipes has been used.

Community Participation

Community participation in the planning of any sewer system is a fundamental requirement to achieve higher household connection rates and to increase the likelihood of proper maintenance of in-block sewers. In addition, it can motivate users to assume parts of the costs of the sewer system that they are able to assume, such as contribution of labor for construction and/or maintenance.

Typically, in the planning process for a simplified sewerage system, meetings are carried out at the housing block (*condominio*) level for information, discussions and clarifications required for a joint group decision on network design, community contributions during construction and maintenance responsibilities. Users might finance and implement in-house sanitary installations and household connections and would agree on a suitable type of condominial branch. They are asked to comply with agreements established for construction and operation of the condominial branch, as well as payment of tariffs. In turn, the service provider agrees to fulfill his responsibilities as established in the "Terms of Connection " between the parties. The community participation process also provides a good opportunity for complementary actions like hygiene promotion, which can have a significant impact on public health at a relatively limited cost.

Design and Construction

Simplified sewers are usually laid in the front yard or under the pavement (sidewalk). In some rare cases it is possible to lay them in the back yard. Sidewalk branches are usually preferred in regular urbanizations, while the front and back yard branches are particularly suited to neighborhoods with challenging topography or urbanization patterns. However, in some cases neither of these options is possible. For example, in South Asia, in many cities there is no sidewalk or front yard, so pipes have to be laid in the middle of the street as with conventional sewers.

In Latin America typical simplified sewer diameters are 100 mm, laid at a gradient of 1 in 200 (0.5 percent). Such a sewer will serve around 200 households of 5 people with a wastewater flow of 80 litres per person per day. In Pakistan, however, there are no rigorous standards for sewer diameters. In a small pilot as part of the Orangi Pilot Project pipes with a diameter of 150mm were used.

Laying small diameter pipes at fairly flat gradients requires careful construction techniques. Plastic pipes are best used as they are more easily jointed correctly. This reduces wastewater leakage from the sewer and groundwater infiltration into it. With simplified sewerage there is no need to have the large expensive manholes of the type used for conventional sewerage — simple brick or plastic junction chambers are used instead.

Construction can be carried out by contractors or by trained and properly supervised community members. Training and proper supervision are actually needed in both cases, since contractors in many cities are not familiar with simplified sewerage.

Investment Cost Comparison

The cost of sewerage - conventional or simplified - are always site-specific, and estimates are subject to controversies. Construction costs of simplified sewerage are *up to* half the costs of conventional sewerage. Investment cost savings come from various design features that may or may not be present in a particular simplified sewerage system. Cost-saving features of any simplified sewerage system are a smaller diameter of pipes, smaller and shallower trenches and simplified manholes. The two latter features are estimated to account for most of the cost savings. Other features that could further reduce costs may only be present in some systems, such as:

- shorter networks;
- avoidance of the need to damage pavements and sidewalks (if they already exist and if pipes are laid in front or back yards);
- decentralized, small-scale wastewater treatment, and consequently elimination of main collectors and sewage pumping stations.

An element that may slightly increase costs compared to conventional sewerage is the introduction of grease traps or of interceptor chambers for the settlement of sludge. The latter are more common in South Asia and are not used in the condominial model. A 2006 study of four countries showed cost savings of 31-57% from the use of simplified sewerage compared to conventional sewerage with unit costs varying from US$119 per connection in a neighborhood in Bolivia and to US$759 per connection in a small town in Paraguay. A detailed estimate gives the costs of simplified sewerage in Lima as at least US$700 per household (US$120–140 per person), including in-house sanitary facilities (US$100 per household) and including design, supervision and social intermediation costs (US$126 per household, which are common costs shared with water infrastructure), but excluding taxes.

In general, at higher population densities sewer systems are cheaper than on-site sanitation (such as septic tanks). The switching value at which sewerage becomes less costly is largely determined by the type of sewerage, conventional or simplified. A 1983 study in Natal showed that the investment costs for simplified sewerage were lower than for on-site systems at the quite low population density of about 160 people per hectare. Conventional sewerage, however, was cheaper only at densities above 400 people per hectare.

Operation and Maintenance

Good operation and maintenance (O&M) is essential for the long-term sustainability of any sewerage system, but particularly for simplified sewerage, since the low diameter of pipes and low gradients make the system more vulnerable to clogging.

The original concept of householders being responsible for O&M of in-block condominial sewers has not worked well in the long term. A study of simplified sewerage systems in Brazil has shown that effective maintenance of sewers by utilities has often been the result of community pressure by neighborhood associations. Without such pressure maintenance by utilities has often been inadequate, and community maintenance has not come about either.

Therefore, alternative management systems had to be developed and a few examples are provided below:

- In rural Ceará a villager is employed by the Residents' Association to maintain the sewers and the wastewater treatment plant (typically, a single facultative waste stabilization pond). He is also responsible for the water supply.
- In parts of Recife in northeast Brazil the state water and sewerage company employs local contracting firms for O&M. Usually this is done by a small team comprising a technician engineer and two laborers who work in a low-income area served by simplified sewerage and to whom residents report any problems.
- In Brasília the water and sewerage company, which has over 1,200 km of condominial sewers, uses van-mounted water jet units to clear any blockages.

Concerning maintenance costs, available information indicates similar costs and requirements for the simplified and the conventional system under the same conditions. Simplified systems typically require more interventions, but the cost per intervention is lower. Comparative analytical studies are not yet available, however.

Constraints for Application

According to Jose Carlos Melo, who is considered to be the "father" of condominial sewers in Brazil, some important constraints for the application of simplified sewerage are:

- Lack of information on fundamentals and techniques of the approach or lack of experience in its application,
- Resistance to change: Institutional, technical and operational changes required by the service provider for implementing the condominial approach usually provoke resistance and can hinder the application.
- Normative and legal restrictions: Existing conservative design and construction standards linked to conventional systems can be an essential constraint in the introduction and dissemination of the systems.

Over the last years, countries like Bolivia and Peru reviewed and modernized technical standards according to methods and criteria established and accepted in Brazil in the 1980s, thus overcoming the latter constraint.

Advantages and Disadvantages of Combined System

Advantages

- In an area where rainfall is spread throughout a year, there is no need of flushing of sewers, as self cleansing velocity will be developed due to more quantity because of addition of storm water.

- Only one set of pipe will be required for house plumbing.
- In congested areas it is easy to lay only one pipe rather than two pipes as required in other systems.

Disadvantages

- Not suitable for the area with small period of rainfall in a year, because dry weather flow will be small due to which self cleansing velocity may not develop in sewers, resulting in silting.
- Large flow is required to be treated at sewage treatment plant before disposal, hence resulting in higher capital and operating cost of the treatment plant.
- When pumping is required this system is uneconomical.
- During rains overflowing of sewers will spoil public hygiene.

Advantages and Disadvantages of Separate System Advantages

Advantages

- As sewage flows in separate pipe, hence the quantity to be treated at sewage treatment plant is small, resulting in economy of treatment.
- This system may be less costly as only sanitary sewage is transported in closed conduit and storm water can be collected and conveyed through open drains.
- When pumping is required during disposal, this system is economical due to less flow.

Disadvantages

- Self cleansing velocity may not developed at certain locations in sewers and hence flushing of sewers may be required.
- This system requires laying two sets of pipe, which may be difficult in congested area.
- This system will require maintenance of two sets of pipelines and hence maintenance cost is more.

Advantages and Disadvantages of Partially Separate System

Advantages

- Economical and reasonable size sewers are required.
- Work of house plumbing is reduced as rain water from roofs, sullage from bathrooms and kitchen, etc. are combined with discharge from water closets.
- Flushing of sewers may not be required as small portion of storm water is allowed to enter in sanitary sewage.

Disadvantages

- Increased cost of pumping as compared to separate system at treatment plants and intermediate pumping station wherever required.
- In dry weather self-cleansing velocity may not develop in the sewers.

Considerations for the Type of System

Following points are considered before finalizing the type of collection system:

- The separate system requires laying of two sets of conduits whereas in combined system only one bigger size conduit is required.
- Laying of two separate conduits may be difficult in the congested streets.
- In combined system sewers are liable for silting during non-monsoon season, hence they are required to be laid at steeper gradients. Steeper gradients for the sewers may require more number of pumping stations, particularly for flat terrain, which may make the system costly.
- Large quantity of wastewater is required to be treated before discharge in case of combined system. Hence, large capacity treatment plant is required.
- In separate system, only sewage is treated before it is discharged into natural water body or used for irrigation. No treatment is generally given to the rainwater collected before it is discharge in to natural water body.
- In case of separate system pumping is only required for sewage. Pumping can be avoided for storm water lines, as they are not very deep and normally laid along the natural slopes.

  In combined system large capacity pumping station is required to safely handle the flow that is likely to be generated during highest design storm considered.
- Based on site conditions the economy of the system needs to be evaluated and selection is made accordingly.

Patterns of Collection System

The network of sewers consists of house sewers discharging the sewage to laterals. The lateral discharges the sewage into branch sewers or sub-mains and sub-mains discharge it into main sewer or trunk sewer. The trunk sewer carries sewage to the common point where adequate treatment is given to the sewage and then it is discharged. The patterns of collection system depend upon:

- The topographical and hydrological features of the area.
- The location and methods of treatment and disposal works.

- ➢ The type of sewerage system employed,
- ➢ Extent of area to be served.

Following patterns can be adopted for collection systems as per the suitability (Birdie, 1990).

Perpendicular Pattern

- ➢ The shortest possible path is maintained for the rains carrying storm water and sewage.
- ➢ It is suitable for separate system and partially separate system for storm water drains.
- ➢ This pattern is not suitable for combined system, because treatment plant is required to be installed at many places; otherwise it will pollute the water body where the sewage is discharged.

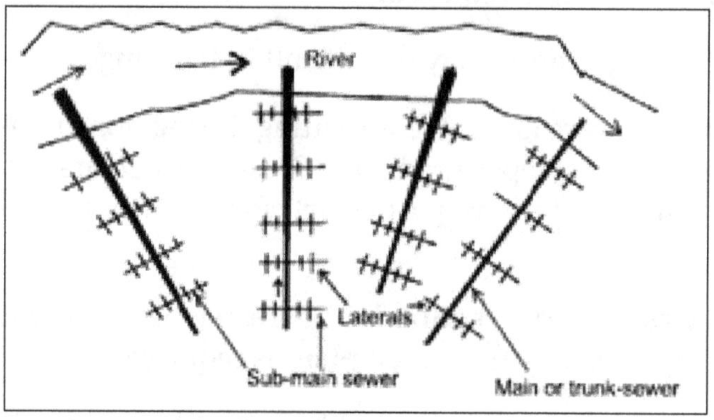

Perpendicular pattern of collection system.

Interceptor Pattern

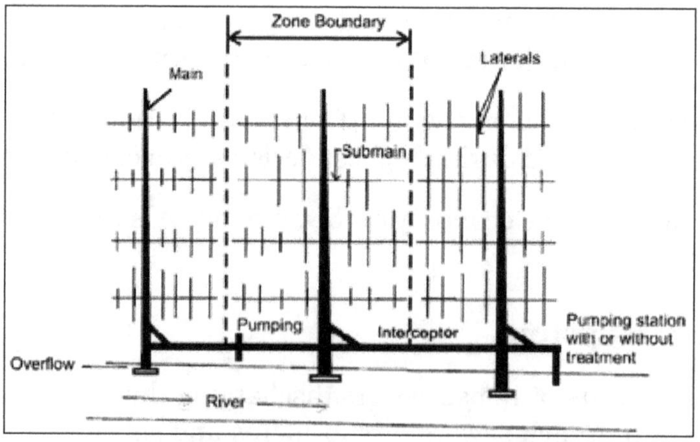

Interceptor pattern of collection system.

- ➢ Sewers are intercepted with large size sewers.
- ➢ Interceptor carries sewage to a common point, where it can be disposed off with or without treatment.
- ➢ Overflows should be provided to handle very large flow.

Radial Pattern

- It is suitable for land disposal.
- In this pattern sewers are laid radialy outwards from the centre, hence this pattern is called as radial pattern.
- The drawback in this pattern is more number of disposal works are required.

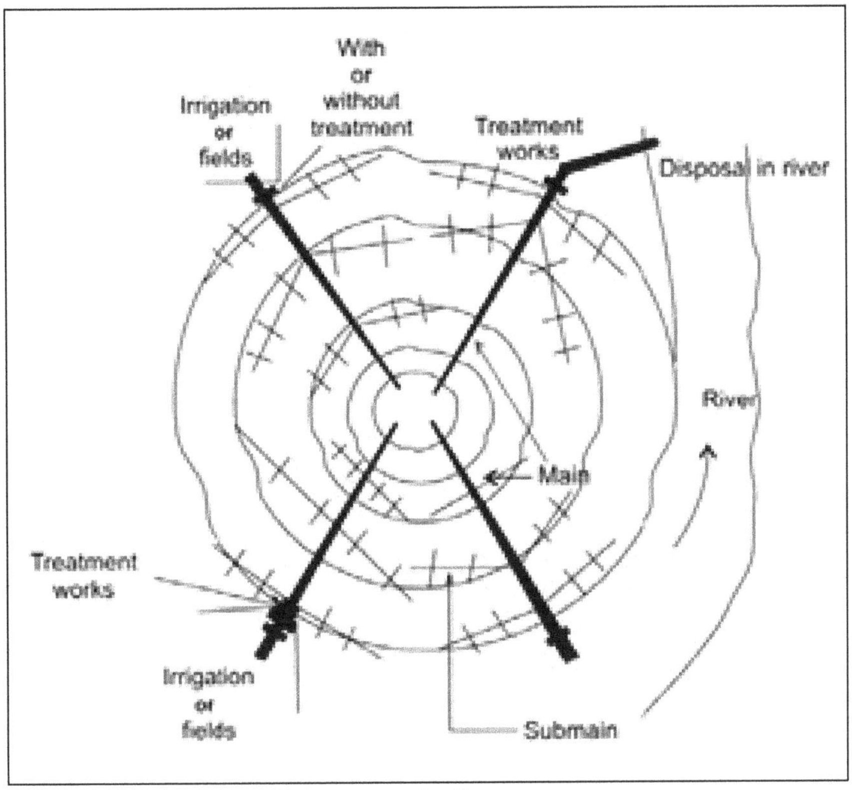

Radial pattern of collection system.

Fan Pattern

- This pattern is suitable for a city situated at one side of the natural water body, such as river.
- The entire sewage flows to a common point where one treatment plant is located.
- In this number of converging main sewers and sub-mains are used forming a fan shape.
- Single treatment plant is required in this pattern.
- The drawback in this pattern is that larger diameter sewer is required near to the treatment plant as entire sewage is collected at a common point.
- In addition, with new development of the city the load on existing treatment plant increases.

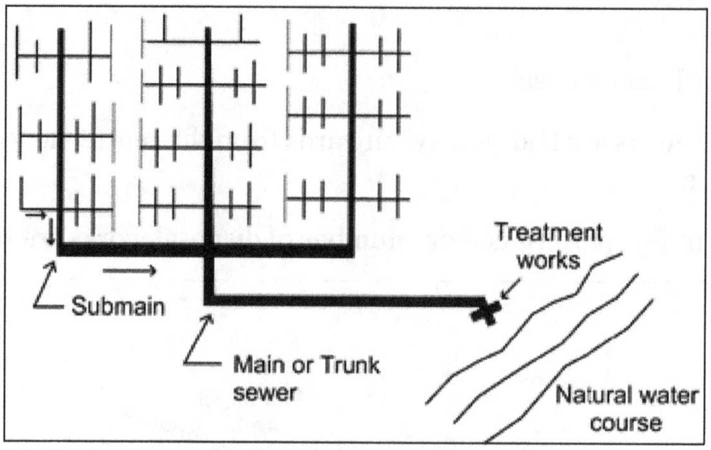

Fan pattern of collection system.

Zone Pattern

- More numbers of interceptors are provided in this pattern.
- This pattern is suitable for sloping area than flat areas.

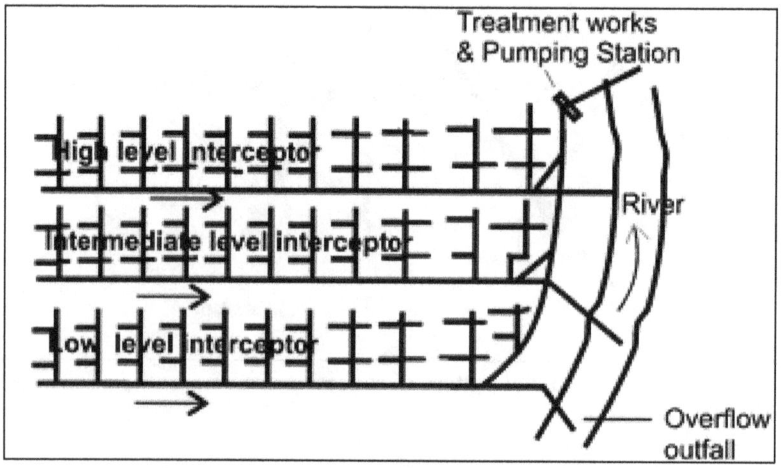

Zone pattern of collection system.

Quantity Estimation of Sewage

The sewage collected from the municipal area consists of wastewater generated from the residences, commercial centers, recreational activities, institutions and industrial wastewaters discharge into sewer network from the permissible industries located within the city limits. Before designing the sewer, it is necessary to know the discharge i.e., quantity of sewage, which will flow in it after completion of the project.

Accurate estimation of sewage discharge is necessary for hydraulic design of the sewers. Far lower estimation than reality will soon lead to inadequate sewer size after commissioning of the scheme or the sewers may not remain adequate for the entire design period. Similarly, very high discharge estimated will lead to larger sewer size affecting economy of the sewerage scheme, and the lower discharge actually flowing in the sewer may not meet the criteria of the self cleansing velocity and hence leading to deposition in the sewers. Actual measurement of the discharge is not possible if the sewers do not exist; and where the capacity of the existing sewers is inadequate and need to be increased, still actual present discharge measurement may not be accurate due to unaccounted overflow and leakages that might be occurring in the existing system. Since sewers are design to serve for some more future years, engineering skills have to be used to accurately estimate the sewage discharge.

Sources of Sanitary Sewage

1. Water supplied by water authority for domestic usage, after desired use it is discharged in to sewers as sewage.
2. Water supplied to the various industries for various industrial processes by local authority. Some quantity of this water after use in different industrial applications is discharged as wastewater.
3. The water supplied to the various public places such as, schools, cinema theaters, hotels, hospitals, and commercial complexes. Part of this water after desired use joins the sewers as wastewater.
4. Water drawn from wells by individuals to fulfill domestic demand. After uses this water is discharged in to sewers.
5. The water drawn for various purposes by industries, from individual water sources such as, wells, tube wells, lake, river, etc. Fraction of this water is converted into wastewater in different industrial processes or used for public utilities within the industry generating wastewater. This is discharged in to sewers.
6. Infiltration of groundwater into sewers through leaky joints.
7. Entrance of rainwater in sewers during rainy season through faulty joints or cracks in sewers.

Dry Weather Flow

Dry weather flow is the flow that occurs in sewers in separate sewerage system or the flow that occurs during dry seasons in combined system. This flow indicates the flow of sanitary sewage. This depends upon the rate of water supply, type of area served, economic conditions of the people, weather conditions and infiltration of groundwater in the sewers, if sewers are laid below groundwater table.

Evaluation of Sewage Discharge

Correct estimation of sewage discharge is necessary; otherwise sewers may prove inadequate resulting in overflow or may prove too large in diameter, which may make the system uneconomical and hydraulically inefficient. Hence, before designing the sewerage system it is important to know the discharge / quantity of the sewage, which will flow in it after completion of the project and at the end of design period.

Apart from accounted water supplied by water authority that will be converted to wastewater, following quantities are considered while estimating the sewage quantity:

Addition Due to Unaccounted Private Water Supplies

People using water supply from private wells, tube wells, etc. contribute to the wastewater generation more than the water supplied by municipal authority. Similarly, certain industries utilize their own source of water. Part of this water, after desired uses, is converted into wastewater and ultimately discharged into sewers. This quantity can be estimated by actual field observations.

Addition Due to Infiltration

This is additional quantity due to groundwater seepage in to sewers through faulty joints or cracks formed in the pipes. The quantity of the water depends upon the height of the water table above the sewer invert level. If water table is well below the sewer invert level, the infiltration can occur only after rain when water is moving down through soil. Quantity of the water entering in sewers depends upon the permeability of the ground soil and it is very difficult to estimate. While estimating the design discharge, following suggested discharge can be considered.

Table Suggested estimates for groundwater infiltration for sewers laid below groundwater table (CPHEEO Manual, 1993).

Unit	Minimum	Maximum
L/ha.d	5000	50000
L/km.d	500	5000
L per day per manhole	250	500

Storm water drainage may also infiltrate into sewers. This inflow is difficult to calculate. Generally, no extra provision is made for this quantity. This extra quantity can be taken care of by extra empty space left at the top in the sewers, which are designed for running ¾ full at maximum design discharge.

Subtraction Due to Water Losses

The water loss, through leakage in water distribution system and house connections, does not reach consumers and hence, not appear as sewage.

Subtraction Due to Water not Entering the Sewerage System

Certain amount of water is used for such purposes, which may not generate sewage, e.g. boiler feed water, water sprinkled over the roads, streets, lawns, and gardens, water consumed in industrial product, water used in air coolers, etc.

Net quantity of sewage: The net quantity of sewage production can be estimated by considering the addition and subtraction as discussed above over the accounted quantity of water supplied by water authority as below:

$$\text{Net quantity of sewage} = \begin{array}{l}\text{Accounted}\\ \text{quantity of}\\ \text{water supplied,}\\ \text{from the}\\ \text{waterworks}\end{array} \begin{array}{l}\text{Addition due to}\\ \text{unaccounted}\\ \text{private water}\\ \text{supplies}\end{array} + \begin{array}{l}\text{Additiondue}\\ \text{to infiltration}\end{array} - \begin{array}{l}\text{Subtraction}\\ \text{due to water}\\ \text{not entering}\\ \text{the sewerage}\\ \text{system}\end{array}$$

Generally, 75 to 80% of accounted water supplied is considered as quantity of sewage produced.

Variation in Sewage Flow

Variation occurs in the flow of sewage over annual average daily flow. Fluctuation in flow occurs from hour to hour and from season to season. The typical hourly variation in the sewage flow is shown in the figure below. If the flow is gauged near its origin, the peak flow will be quite pronounced. The peak will defer if the sewage has to travel long distance. This is because of the time required in collecting sufficient quantity of sewage required to fill the sewers and time required in travelling. As sewage flow in sewer lines, more and more sewage is mixed in it due to continuous increase in the area being served by the sewer line. This leads to reduction in the fluctuations in the sewage flow and the lag period goes on increasing. The magnitude of variation in the sewage quantity varies from place to place and it is very difficult to predict. For smaller township this variation will be more pronounced due to lower length and travel time before sewage reach to the main sewer and for large cities this variation will be less.

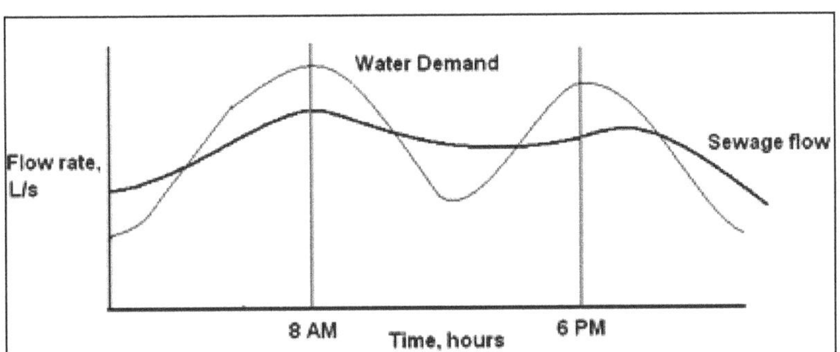

Typical hourly variations in sewage flow.

For estimating design discharge following relation can be considered:

Maximum daily flow = Two times the annual average daily flow (representing seasonal variations)

Maximum hourly flow= 1.5 times the maximum daily flow (accounting hourly variations)

= Three times the annual average daily flow

As the tributary area increases, peak hourly flow will decrease. For smaller population served (less than 50000) the peak factor can be 2.5, and as the population served increases its value reduces. For large cities it can be considered about 1.5 to 2.0. Therefore, for outfall sewer the peak flow can be considered as 1.5 times the annual average daily flow. Even for design of the treatment facility, the peak factor is considered as 1.5 times the annual average daily flow.

The minimum flow passing through sewers is also important to develop self cleansing velocity to avoid silting in sewers. This flow will generate in the sewers during late night hours. The effect of this flow is more pronounced on lateral sewers than the main sewers. Sewers must be checked for minimum velocity as follows:

Minimum daily flow = 2/3 Annual average daily flow

Minimum hourly flow = ½ minimum daily flow

= 1/3 Annual average daily flow

The overall variation between the maximum and minimum flow is more in the laterals and less in the main or trunk sewers. This ratio may be more than 6 for laterals and about 2 to 3 in case of main sewers.

Design Period

The future period for which the provision is made in designing the capacities of the various components of the sewerage scheme is known as the design period. The design period depends upon the following:

- Ease and difficulty in expansion,
- Amount and availability of investment,
- Anticipated rate of population growth, including shifts in communities, industries and commercial investments,
- Hydraulic constraints of the systems designed,
- Life of the material and equipment.

Following design period can be considered for different components of sewerage scheme.

1. Laterals less than 15 cm diameter : Full development
2. Trunk or main sewers : 40 to 50 years
3. Treatment Units : 15 to 20 years
4. Pumping plant : 5 to 10 years

Design Discharge of Sanitary Sewage

The total quantity of sewage generated per day is estimated as product of forecasted population at the end of design period considering per capita sewage generation and appropriate peak factor. The per capita sewage generation can be considered as 75 to 80% of the per capita water supplied per day. The increase in population also result in increase in per capita water demand and hence, per capita production of sewage. This increase in water demand occurs due to increase in living standards, betterment in economical condition, changes in habit of people, and enhanced demand for public utilities.

Quantity Estimation of Storm Water

The surface run-off resulting after precipitation contributes to the stormwater. The quantity of stormwater reaching to the sewers or drains is very large as compared with sanitary sewage. The factors affecting the quantity of stormwater flow are as below:

- Area of the catchment.
- Slope and shape of the catchment area.
- Porosity of the soil.
- Obstruction in the flow of water as trees, fields, gardens, etc.
- Initial state of catchment area with respect to wetness.
- Intensity and duration of rainfall.
- Atmospheric temperature and humidity.
- Number and size of ditches present in the area.

Measurement of Rainfall

The rainfall intensity could be measured by using rain gauges and recording the amount of rain falling in unit time. The rainfall intensity is usually expressed as mm/hour or cm/hour. The rain gauges used can be manual recording type or automatic recording rain gauges.

Methods for Estimation of Quantity of Storm Water

- Rational Method
- Empirical formulae method

In both the above methods, the quantity of storm water is considered as function of intensity of rainfall, coefficient of runoff and area of catchment.

Time of Concentration: The period after which the entire catchment area will start contributing to the runoff is called as the time of concentration.

- The rainfall with duration lesser than the time of concentration will not produce maximum discharge.
- The runoff may not be maximum even when the duration of the rain is more than the time of concentration. This is because in such cases the intensity of rain reduces with the increase in its duration.

- The runoff will be maximum when the duration of rainfall is equal to the time of concentration and is called as critical rainfall duration. The time of concentration is equal to sum of inlet time and time of travel.

Time of concentration = Inlet time + time of travel

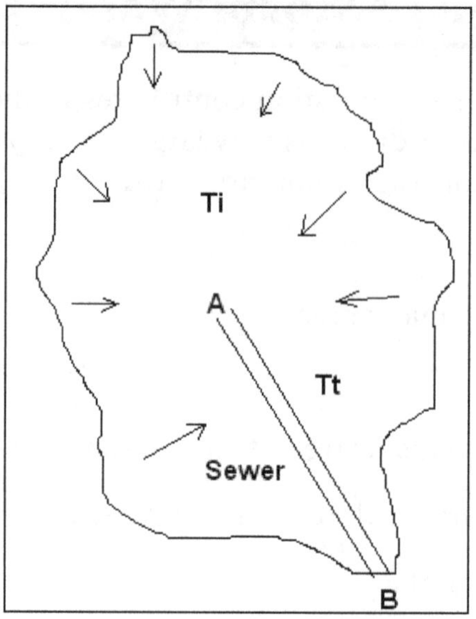

Runoff from a given catchment.

Inlet Time: The time required for the rain in falling on the most remote point of the tributary area to flow across the ground surface along the natural drains or gutters up to inlet of sewer is called inlet time. The inlet time 'Ti' can be estimated using relationships similar to following. These coefficients will have different values for different catchments.

$$Ti = [0.885\ L^3/H]^{0.385}$$

Where,

- Ti = Time of inlet, minute
- L = Length of overland flow in Kilometer from critical point to mouth of drain
- H = Total fall of level from the critical point to mouth of drain, meter

Time of Travel: The time required by the water to flow in the drain channel from the mouth to the point under consideration or the point of concentration is called as time of travel.

Time of Travel (Tt) = Length of drain/ velocity in drain

Runoff Coefficient: The total precipitation falling on any area is dispersed as percolation, evaporation, storage in ponds or reservoir and surface runoff. The runoff coefficient can be defined as a fraction, which is multiplied with the quantity of total rainfall to determine the quantity of rain water, which will reach the sewers. The runoff coefficient depends upon the porosity of soil cover, wetness and ground cover. The overall runoff coefficient for the catchment area can be worked out as follows:

Overall runoff coefficient, $C = [A_1.C1 + A_2.C_2 + ... + A_n.C_n] / [A_1 + A_2 + ... + A_n]$

Where, A_1, A_2, ...An are types of area with C_1, C_2, ...C_n as their coefficient of runoff, respectively.

The typical runoff coefficient for the different ground cover is provided in the table below. The Table Runoff coefficient for different type of cover in catchment.

Type of Cover	Coefficient of runoff
Business areas	0.70 – 0.90
Apartment areas	0.50 – 0.70
Single family area	0.30 – 0.50
Parks, Playgrounds, Lawns	0.10 – 0.25
Paved Streets	0.80 –0.90
Water tight roofs	0.70 – 0.95

Rational Method

Storm water quantity can be estimated by rational method as below:

Storm water quantity, Q = C.I.A / 360

Where,

- Q = Quantity of storm water, m^3/sec
- C = Coefficient of runoff
- I = intensity of rainfall (mm/hour) for the duration equal to time of concentration, and
- A = Drainage area in hectares

OR

- Q = 0.278 C.I.A

Where, Q is m^3/sec; I is mm/hour, and A is area in square kilometer

Empirical Formulae

Empirical formulae are used for determination of runoff from very large area. Various empirical relationships are developed based on the past observations on specific site conditions suiting a particular region. These empirical formulae can be used for prediction of storm water runoff for that particular catchment.

A] Burkli – Zeiglar formula

$$Q = \frac{C.I.A}{141.58} \sqrt[4]{S/A}$$

B] Mc Math formula (used in USA)

$$Q = \frac{C.I.A}{148.35} \sqrt[5]{S/A}$$

C] Fuller's formula

$$Q = \frac{C.M^{0.8}}{13.23}$$

(Where, S- Slope of the area in meter per thousand meter, M- drainage area in sq. km., A – drainage area in hectare)

Empirical Formulae for Rainfall Intensities

These relationships between rainfall intensity and duration are developed based on long term experience in field. Under Indian conditions, intensity of rainfall in design is usually in the range 12 mm/h to 20 mm/h. In general, the empirical relationship has the following forms:

$$I = a/(t + b) \quad \text{OR} \quad I = b/t^n$$

Where, a, b, and n are constants.

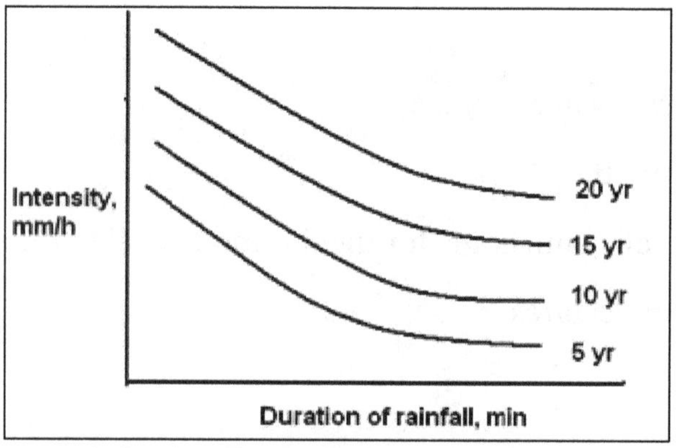

Relationship of rainfall duration and intensity

British Ministry of Health formula

$I = 760 / (t + 10)$ (for storm duration of 5 to 20 minutes)

$I = 1020 / (t + 10)$ (for storm duration of 20 to 100 minutes)

Where, I is intensity of rainfall, mm/h and t is duration of storm, minutes.

References

- Benidickson, Jamie (2011). The Culture of Flushing: A Social and Legal History of Sewage. UBC Press. ISBN 9780774841382. Retrieved 15, March 2020
- "Going Against the Flow: Green Tech, Sensors and Industrial Internet Make Sewer Systems Smart". Txchnologist. General Electric. 2013. Retrieved 28, February 2020

- L. Montestruque, M. Lemmon (2015). "Globally Coordinated Distributed Storm Water Management System." 1st International Workshop on Cyber Physical Systems for Smart Water Networks, doi:10.1145/2738935.2738948
- Fan, Chi-Yuan; Field, Richard; Lai, Fu-hsiung. "Sewer-Sediment Control: Overview of an EPA Wet-Weather Flow Research Program" (PDF). University of California Los Angeles. United States Environmental Protection Agency. Retrieved 05, July 2020
- Tilley, E., Ulrich, L., Lüthi, C., Reymond, Ph., Zurbrügg, C. (2014) Compendium of Sanitation Systems and Technologies - (2nd Revised Edition). Swiss Federal Institute of Aquatic Science and Technology (Eawag), Duebendorf, Switzerland. ISBN 978-3-906484-57-0
- Dhiman, Gaurav (2016). "Using Multiple Antibiotic Resistance Profiles of Coliforms as a Tool to Investigate Combined Sewer Overflow Contamination". Journal of Environmental Health

Chapter 6

Diverse Methods for Wastewater Treatment

Wastewater treatment is defined as the removal of impurities and contaminants from wastewater or sewage before it is released back into the environment. Wastewater can be treated through a number of physical, chemical and biological processes or any combination of the three. Sedimentation wastewater treatment, sewage sludge treatment, bar screen, etc. are some techniques which are employed for the treatment of wastewater. This chapter is a comprehensive summary of these wastewater treatment techniques.

Wastewater Treatment

Wastewater treatment is the process of converting wastewater – water that is no longer needed or is no longer suitable for use – into bilge water that can be discharged back into the environment. It's formed by a number of activities including bathing, washing, using the toilet, and rainwater runoff. Wastewater is full of contaminants including bacteria, chemicals and other toxins. Its treatment aims at reducing the contaminants to acceptable levels to make the water safe for discharge back into the environment.

There are two wastewater treatment plants namely chemical or physical treatment plant, and biological wastewater treatment plant. Biological waste treatment plants use biological matter and bacteria to break down waste matter. Physical waste treatment plants use chemical reactions as well as physical processes to treat wastewater. Biological treatment systems are ideal for treating wastewater from households and business premises. Physical wastewater treatment plants are mostly used to treat wastewater from industries, factories and manufacturing firms. This is because most of the wastewater from these industries contains chemicals and other toxins that can largely harm the environment.

Step by Step Wastewater Treatment Process

The following is a step by step process of how wastewater is treated:

Wastewater Collection

This is the first step in waste water treatment process. Collection systems are put in place by municipal administration, home owners as well as business owners to ensure that all the wastewater is collected and directed to a central point. This water is then directed to a treatment plant using underground drainage systems or by exhauster tracks owned and operated by business people. The transportation of wastewater should however be done under hygienic conditions. The pipes or tracks should be leak proof and the people offering the exhausting services should wear protective clothing.

Odor Control

At the treatment plant, odor control is very important. Wastewater contains a lot of dirty substances that cause a foul smell over time. To ensure that the surrounding areas are free of the foul smell, odor treatment processes are initiated at the treatment plant. All odor sources are contained and treated using chemicals to neutralize the foul smell producing elements. It is the first wastewater treatment plant process and it's very important.

Screening

This is the next step in wastewater treatment process. Screening involves the removal of large objects for example nappies, cotton buds, plastics, diapers, rags, sanitary items, nappies, face wipes, broken bottles or bottle tops that in one way or another may damage the equipment. Failure to observe this step, results in constant machine and equipment problems. Specially designed equipment is used to get rid of grit that is usually washed down into the sewer lines by rainwater. The solid wastes removed from the wastewater are then transported and disposed off in landfills.

Primary Treatment

This process involves the separation of macrobiotic solid matter from the wastewater. Primary treatment is done by pouring the wastewater into big tanks for the solid matter to settle at the surface of the tanks. The sludge, the solid waste that settles at the surface of the tanks, is removed by large scrappers and is pushed to the center of the cylindrical tanks and later pumped out of the tanks for further treatment. The remaining water is then pumped for secondary treatment.

Secondary Treatment

Also known as the activated sludge process, the secondary treatment stage involves adding seed sludge to the wastewater to ensure that is broken down further. Air is first pumped into huge aeration tanks which mix the wastewater with the seed sludge which is basically small amount of sludge, which fuels the growth of bacteria that uses oxygen and the growth of other small microorganisms that consume the remaining organic matter. This process leads to the production of large particles that settle down at the bottom of the huge tanks. The wastewater passes through the large tanks for a period of 3-6 hours.

Bio-solids Handling

The solid matter that settle out after the primary and secondary treatment stages are directed to digesters. The digesters are heated at room temperature. The solid wastes are then treated for a month where they undergo anaerobic digestion. During this process, methane gases are produced and there is a formation of nutrient rich bio-solids which are recycled and dewatered into local firms. The methane gas formed is usually used as a source of energy at the treatment plants. It can be used to produce electricity in engines or to simply drive plant equipment. This gas can also be used in boilers to generate heat for digesters.

Tertiary Treatment

This stage is similar to the one used by drinking water treatment plants which clean raw water for drinking purposes. The tertiary treatment stage has the ability to remove up to 99 percent of the impurities from the wastewater. This produces effluent water that is close to drinking water quality. Unfortunately, this process tends to be a bit expensive as it requires special equipment, well trained and highly skilled equipment operators, chemicals and a steady energy supply. All these are not readily available.

Disinfection

After the primary treatment stage and the secondary treatment process, there are still some diseases causing organisms in the remaining treated wastewater. To eliminate them, the wastewater must be disinfected for at least 20-25 minutes in tanks that contain a mixture of chlorine and sodium hypochlorite. The disinfection process is an integral part of the treatment process because it guards the health of the animals and the local people who use the water for other purposes. The effluent (treated waste water) is later released into the environment through the local water ways.

Sludge Treatment

The sludge that is produced and collected during the primary and secondary treatment processes requires concentration and thickening to enable further processing. It is put into thickening tanks that allow it to settle down and later separates from the water. This process can take up to 24 hours. The remaining water is collected and sent back to the huge aeration tanks for further treatment. The sludge is then treated and sent back into the environment and can be used for agricultural use.

Industrial Wastewater Treatment

Industrial wastewater treatment is applicable to most manufacturing, mining, energy and petrochemical companies aiming to decrease their liquid discharge into the environment. This benefits the environment and, importantly, helps companies save money by re-using large water volumes in their industrial processes.

The Potential of Industrial Wastewater Treatment

Environmental compliance

Veolia aims to help industrial clients comply with region-specific regulations pertaining to industrial wastewater treatment standards.

Process Water Re-use

By recycling water within a closed industrial wastewater treatment system, companies can achieve savings by feeding less clean water into a plant, and can reduce carbon and other pollutants entering the environment. Industrial wastewater treatment and re-use also help to free up potable water for distribution to communities.

Industrial Process Water from Seawater

With industry-leading desalination techniques, Veolia can turn high-salinity seawater into suitable for use in industrial processes. It is a well-established process of removing salts from water to produce process water using technologies like reverse osmosis, multiple effect distillation, multi-stage flash and hybrid systems that use thermal and reverse osmosis.

Industrial Wastewater Treatment for Human Consumption

Cutting-edge technologies are able to transform industrial wastewater into a fresh water resource for human consumption through various water purification technologies. Veolia uses leading water purification technologies and, worldwide, Veolia's global company has built more than 5 000 water purification plants.

Value Recovery from Industrial Wastewater

Industrial wastewater treatment technologies are able to extract valuable substances from process water before discharge or re-use within a plant, achieving greater sustainability and lower carbon footprints. By acting sustainably, customers recognise the value of water, help to save energy and produce it, and optimise costs. Reducing your carbon footprint helps create responsible value and a whole lot more.

Sources of Industrial Wastewater

Battery Manufacturing

Battery manufacturers specialize in fabricating small devices for electronics and portable equipment (e.g., power tools), or larger, high-powered units for cars, trucks and other motorized vehicles. Pollutants generated at manufacturing plants includes cadmium, chromium, cobalt, copper, cyanide, iron, lead, manganese, mercury, nickel, oil & grease, silver and zinc.

Organic Chemicals Manufacturing

The specific pollutants discharged by organic chemical manufacturers vary widely from plant to plant, depending on the types of products manufactured, such as bulk organic chemicals, resins, pesticides, plastics, or synthetic fibers. Some of the organic compounds that may be discharged are benzene, chloroform, napthalene, phenols, toluene and vinyl chloride. Biochemical oxygen demand (BOD), which is a gross measurement of a range of organic pollutants, may be used to gauge the effectveness of a biological wastewater treatment system, and is used as a regulatory parameter in some discharge permits. Metal pollutant discharges may include chromium, copper, lead, nickel and zinc.

Electric Power Plants

Fossil-fuel power stations, particularly coal-fired plants, are a major source of industrial wastewater. Many of these plants discharge wastewater with significant levels of metals such as lead, mercury, cadmium and chromium, as well as arsenic, selenium and nitrogen compounds (nitrates and nitrites). Wastewater streams include flue-gas desulfurization, fly ash, bottom ash and flue gas mercury control. Plants with air pollution controls such as wet scrubbers typically transfer the captured pollutants to the wastewater stream.

Ash ponds, a type of surface impoundment, are a widely used treatment technology at coal-fired plants. These ponds use gravity to settle out large particulates (measured as total suspended solids) from power plant wastewater. This technology does not treat dissolved pollutants. Power stations use additional technologies to control pollutants, depending on the particular wastestream in the plant. These include dry ash handling, closed-loop ash recycling, chemical precipitation, biological treatment (such as an activated sludge process), membrane systems, and evaporation-crystallization systems. Technological advancements in ion exchange membranes and electrodialysis systems has enabled high efficiency treatment of flue-gas desulfurization wastewater to meet recent EPA discharge limits. The treatment approach is similar for other highly scaling industrial wastewaters.

Food Industry

Wastewater generated from agricultural and food operations has distinctive characteristics that set it apart from common municipal wastewater managed by public or private sewage treatment plants throughout the world: it is biodegradable and non-toxic, but has high concentrations of biochemical oxygen demand (BOD) and suspended solids (SS). The constituents of food and agriculture wastewater are often complex to predict, due to the differences in BOD and pH in effluents from vegetable, fruit, and meat products and due to the seasonal nature of food processing and post-harvesting.

Processing of food from raw materials requires large volumes of high grade water. Vegetable washing generates waters with high loads of particulate matter and some dissolved organic matter. It may also contain surfactants.

Animal slaughter and processing produces very strong organic waste from body fluids, such as blood, and gut contents. This wastewater is frequently contaminated by significant levels of antibiotics and growth hormones from the animals and by a variety of pesticides used to control external parasites.

Processing food for sale produces wastes generated from cooking which are often rich in plant organic material and may also contain salt, flavourings, colouring material and acids or alkali. Very significant quantities of oil or fats may also be present.

Food processing activities such as plant cleaning, material conveying, bottling, and product washing create wastewater. Many food processing facilities require on-site treatment before operational wastewater can be land applied or discharged to a waterway or a sewer system. High suspended solids levels of organic particles increase Biochemical Oxygen Demand (BOD) can result in significant sewer surcharge fees. Sedimentation, wedgewire screening, or rotating belt filtration (microscreening) are commonly used methods to reduce suspended organic solids loading prior to discharge.

Iron and Steel Industry

The production of iron from its ores involves powerful reduction reactions in blast furnaces. Cooling waters are inevitably contaminated with products especially ammonia and cyanide. Production of coke from coal in coking plants also requires water cooling and the use of water in by-products separation. Contamination of waste streams includes gasification products such as benzene, naphthalene, anthracene, cyanide, ammonia, phenols, cresols together with a range of more complex organic compounds known collectively as polycyclic aromatic hydrocarbons (PAH).

The conversion of iron or steel into sheet, wire or rods requires hot and cold mechanical transformation stages frequently employing water as a lubricant and coolant. Contaminants include hydraulic oils, tallow and particulate solids. Final treatment of iron and steel products before onward sale into manufacturing includes *pickling* in strong mineral acid to remove rust and prepare the surface for tin or chromium plating or for other surface treatments such as galvanisation or painting. The two acids commonly used are hydrochloric acid and sulfuric acid. Wastewaters include acidic rinse waters together with waste acid. Although many plants operate acid recovery plants (particularly those using hydrochloric acid), where the mineral acid is boiled away from the iron salts, there remains a large volume of highly acid ferrous sulfate or ferrous chloride to be disposed of. Many steel industry wastewaters are contaminated by hydraulic oil, also known as *soluble oil.*

Mines and Quarries

Mine wastewater effluent in Peru, with neutralized pH from tailing runoff.

The principal waste-waters associated with mines and quarries are slurries of rock particles in water. These arise from rainfall washing exposed surfaces and haul roads and also from rock washing and grading processes. Volumes of water can be very high, especially rainfall related arisings on large sites. Some specialized separation operations, such as coal washing to separate coal from native rock using density gradients, can produce wastewater contaminated by fine particulate haematite and surfactants. Oils and hydraulic oils are also common contaminants.

Wastewater from metal mines and ore recovery plants are inevitably contaminated by the minerals present in the native rock formations. Following crushing and extraction of the desirable materials, undesirable materials may enter the wastewater stream. For metal mines, this can include unwanted metals such as zinc and other materials such as arsenic. Extraction of high value metals such as gold and silver may generate slimes containing very fine particles in where physical removal of contaminants becomes particularly difficult.

Additionally, the geologic formations that harbour economically valuable metals such as copper and gold very often consist of sulphide-type ores. The processing entails grinding the rock into fine particles and then extracting the desired metal(s), with the leftover rock being known as tailings. These tailings contain a combination of not only undesirable leftover metals, but also sulphide components which eventually form sulphuric acid upon the exposure to air and water that inevitably occurs when the tailings are disposed of in large impoundments. The resulting acid mine drainage, which is often rich in heavy metals (because acids dissolve metals), is one of the many environmental impacts of mining.

Nuclear Industry

The waste production from the nuclear and radio-chemicals industry is dealt with as *Radioactive waste.*

Petroleum Refining and Petrochemicals

Pollutants discharged at petroleum refineries and petrochemical plants include conventional pollutants (biochemical oxygen demand, oil and grease, suspended solids), ammonia, chromium, phenols and sulfides.

Pulp and Paper Industry

Effluent from the pulp and paper industry is generally high in suspended solids and BOD. Plants that bleach wood pulp for paper making may generate chloroform, dioxins (including 2,3,7,8-TCDD), furans, phenols and chemical oxygen demand (COD). Stand-alone paper mills using imported pulp may only require simple primary treatment, such as sedimentation or dissolved air flotation. Increased BOD or COD loadings, as well as organic pollutants, may require biological treatment such as activated sludge or upflow anaerobic sludge blanket reactors. For mills with high inorganic loadings like salt, tertiary treatments may be required, either general membrane treatments like ultrafiltration or reverse osmosis or treatments to remove specific contaminants, such as nutrients.

Textile Dyeing

Textile dyeing plants generate wastewater that contain synthetic and natural dyestuff, gum thickener (guar) and various wetting agents, pH buffers and dye retardants or accelerators. Following treatment with polymer-based flocculants and settling agents, typical monitoring parameters include BOD, COD, color (ADMI), sulfide, oil and grease, phenol, TSS and heavy metals (chromium, zinc, lead, copper).

Industrial Oil Contamination

Industrial applications where oil enters the wastewater stream may include vehicle wash bays, workshops, fuel storage depots, transport hubs and power generation. Often the wastewater is discharged into local sewer or trade waste systems and must meet local environmental specifications. Typical contaminants can include solvents, detergents, grit. lubricants and hydrocarbons.

Water Treatment

Many industries have a need to treat water to obtain very high quality water for demanding purposes such pure chemical synthesis or boiler feed water. Many water treatment produce organic and mineral sludges from filtration and sedimentation. Ion exchange using natural or synthetic resins removes calcium, magnesium and carbonate ions from water, typically replacing them with sodium, chloride, hydroxyl and/or other ions. Regeneration of ion exchange columns with strong acids and alkalis produces a wastewater rich in hardness ions which are readily precipitated out, especially when in admixture with other wastewater constituents.

Wood Preserving

Wood preserving plants generate conventional and toxic pollutants, including arsenic, COD, copper, chromium, abnormally high or low pH, phenols, oil & grease, and suspended solids.

Wool Processing

Insecticide residues in fleeces are a particular problem in treating waters generated in wool processing. Animal fats may be present in the wastewater, which if not contaminated, can be recovered for the production of tallow or further rendering.

Treatment of Industrial Wastewater

The various types of contamination of wastewater require a variety of strategies to remove the contamination.

Brine Treatment

Brine treatment involves removing dissolved salt ions from the waste stream. Although similarities to seawater or brackish water desalination exist, industrial brine treatment may contain unique combinations of dissolved ions, such as hardness ions or other metals, necessitating specific processes and equipment.

Brine treatment systems are typically optimized to either reduce the volume of the final discharge for more economic disposal (as disposal costs are often based on volume) or maximize the recovery of fresh water or salts. Brine treatment systems may also be optimized to reduce electricity consumption, chemical usage, or physical footprint.

Brine treatment is commonly encountered when treating cooling tower blowdown, produced water from steam assisted gravity drainage (SAGD), produced water from natural gas extraction such as coal seam gas, frac flowback water, acid mine or acid rock drainage, reverse osmosis reject, chlor-alkali wastewater, pulp and paper mill effluent, and waste streams from food and beverage processing.

Brine treatment technologies may include: membrane filtration processes, such as reverse osmosis; ion exchange processes such as electrodialysis or weak acid cation exchange; or evaporation processes, such as brine concentrators and crystallizers employing mechanical vapour recompression and steam.

Reverse osmosis may not be viable for brine treatment, due to the potential for fouling caused by hardness salts or organic contaminants, or damage to the reverse osmosis membranes from hydrocarbons.

Evaporation processes are the most widespread for brine treatment as they enable the highest degree of concentration, as high as solid salt. They also produce the highest purity effluent, even distillate-quality. Evaporation processes are also more tolerant of organics, hydrocarbons, or hardness salts. However, energy consumption is high and corrosion may be an issue as the prime mover is concentrated salt water. As a result, evaporation systems typically employ titanium or duplex stainless steel materials.

Brine Management

Brine management examines the broader context of brine treatment and may include consideration of government policy and regulations, corporate sustainability, environmental impact, recycling, handling and transport, containment, centralized compared to on-site treatment, avoidance and reduction, technologies, and economics. Brine management shares some issues with leachate management and more general waste management.

Solids Removal

Most solids can be removed using simple sedimentation techniques with the solids recovered as slurry or sludge. Very fine solids and solids with densities close to the density of water pose special problems. In such case filtration or ultrafiltration may be required. Although, flocculation may be used, using alum salts or the addition of polyelectrolytes. Wastewater from industrial food processing often requires on-site treatment before it can be discharged to prevent or reduce sewer surcharge fees. The type of industry and specific operational practices determine what types of wastewater is generated and what type of treatment is required. Reducing solids such as waste product, organic materials, and sand is often a goal of industrial wastewater treatment. Some common ways to reduce solids include primary sedimentation (clarification), Dissolved Air Flotation or (DAF), belt filtration (microscreening), and drum screening.

Oils and Grease Removal

The effective removal of oils and grease is dependent on the characteristics of the oil in terms of its suspension state and droplet size, which will in turn affect the choice of separator technology. Oil in industrial waste water may be free light oil ,heavy oil, which tends to sink, and emulsified oil, often referred to as soluble oil. Emulsified or soluble ooils will typically required “cracking” to free the oil from its emulsion. In most cases this is achieved by lowering the pH of the water matrix.

Most separator technologies will have an optimum range of oil droplet sizes that can be effectively treated.

Analysing the oily water to determine droplet size can be performed with a video particle analyser. Each separator technology will have its own performance curve outlining optimum performance based on oil droplet size. the most common separators are gravity tanks or pits, API oil-water separators or plate packs, chemical treatment via DAFs, centrifuges, media filters and hydrocyclones.

API Separators

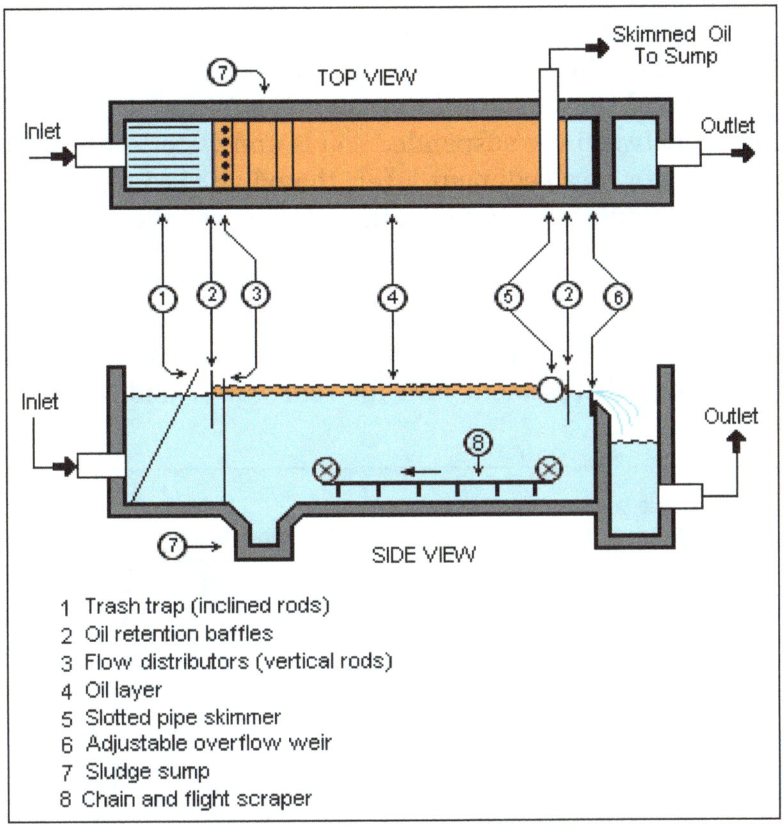

A typical API oil-water separator used in many industries.

Many oils can be recovered from open water surfaces by skimming devices. Considered a dependable and cheap way to remove oil, grease and other hydrocarbons from water, oil skimmers can sometimes achieve the desired level of water purity. At other times, skimming is also a cost-efficient method to remove most of the oil before using membrane filters and chemical processes. Skimmers will prevent filters from blinding prematurely and keep chemical costs down because there is less oil to process.

Because grease skimming involves higher viscosity hydrocarbons, skimmers must be equipped with heaters powerful enough to keep grease fluid for discharge. If floating grease forms into solid clumps or mats, a spray bar, aerator or mechanical apparatus can be used to facilitate removal.

However, hydraulic oils and the majority of oils that have degraded to any extent will also have a soluble or emulsified component that will require further treatment to eliminate. Dissolving or emulsifying oil using surfactants or solvents usually exacerbates the problem rather than solving it, producing wastewater that is more difficult to treat.

The wastewaters from large-scale industries such as oil refineries, petrochemical plants, chemical plants, and natural gas processing plants commonly contain gross amounts of oil and suspended solids. Those industries use a device known as an API oil-water separator which is designed to separate the oil and suspended solids from their wastewater effluents. The name is derived from the fact that such separators are designed according to standards published by the American Petroleum Institute (API).

The API separator is a gravity separation device designed by using Stokes Law to define the rise velocity of oil droplets based on their density and size. The design is based on the specific gravity difference between the oil and the wastewater because that difference is much smaller than the specific gravity difference between the suspended solids and water. The suspended solids settles to the bottom of the separator as a sediment layer, the oil rises to top of the separator and the cleansed wastewater is the middle layer between the oil layer and the solids.

Typically, the oil layer is skimmed off and subsequently re-processed or disposed of, and the bottom sediment layer is removed by a chain and flight scraper (or similar device) and a sludge pump. The water layer is sent to further treatment for additional removal of any residual oil and then to some type of biological treatment unit for removal of undesirable dissolved chemical compounds.

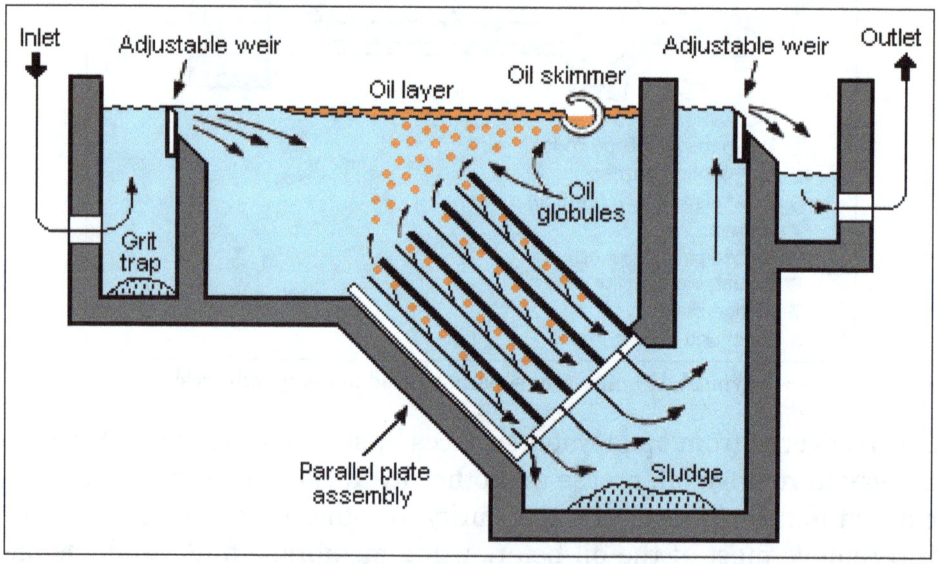

A typical parallel plate separator.

Parallel plate separators are similar to API separators but they include tilted parallel plate assemblies (also known as parallel packs). The parallel plates provide more surface for suspended oil droplets to coalesce into larger globules. Such separators still depend upon the specific gravity between the suspended oil and the water. However, the parallel plates enhance the degree of oil-water separation. The result is that a parallel plate separator requires significantly less space than a conventional API separator to achieve the same degree of separation.

Hydrocyclone Oil Separators

Hydrocyclone oil separators operate on the process where wastewater enters the cyclone chamber and is spun under extreme centrifugal forces more than 1000 times the force of gravity. This force causes the water and oil droplets to separate. The separated oil is discharged from one end of the cyclone where treated water is discharged through the opposite end for further treatment, filtration or discharge.

Removal of Biodegradable Organics

Biodegradable organic material of plant or animal origin is usually possible to treat using extended conventional sewage treatment processes such as activated sludge or trickling filter. Problems can arise if the wastewater is excessively diluted with washing water or is highly concentrated such as undiluted blood or milk. The presence of cleaning agents, disinfectants, pesticides, or antibiotics can have detrimental impacts on treatment processes.

Activated Sludge Process

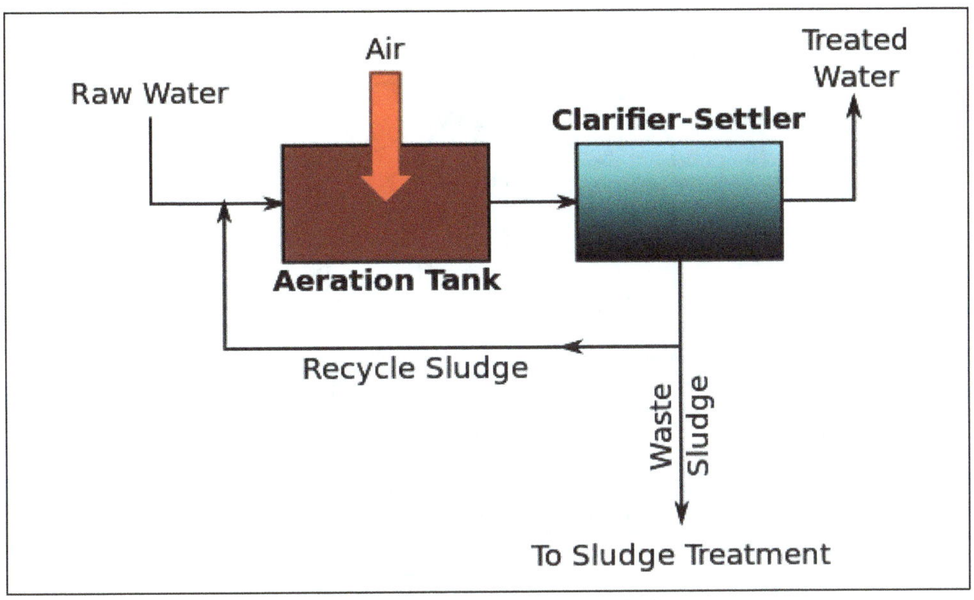

A generalized diagram of an activated sludge process.

Activated sludge is a biochemical process for treating sewage and industrial wastewater that uses air (or oxygen) and microorganisms to biologically oxidize organic pollutants, producing a waste sludge (or floc) containing the oxidized material. In general, an activated sludge process includes:

- An aeration tank where air (or oxygen) is injected and thoroughly mixed into the wastewater.
- A settling tank (usually referred to as a clarifier or "settler") to allow the waste sludge to settle. Part of the waste sludge is recycled to the aeration tank and the remaining waste sludge is removed for further treatment and ultimate disposal.

Trickling Filter Process

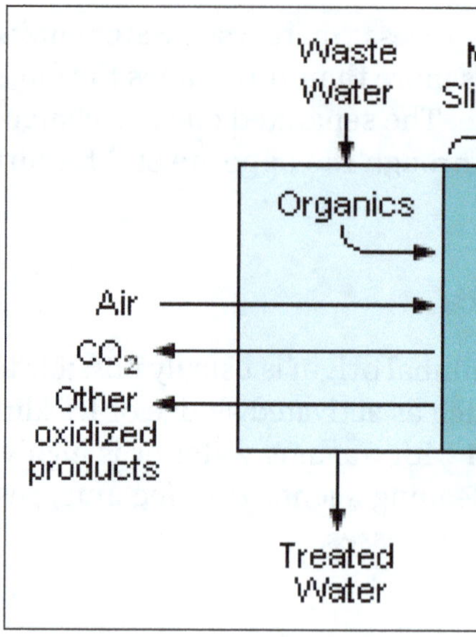

A schematic cross-section of the contact face of the bed media in a trickling filter.

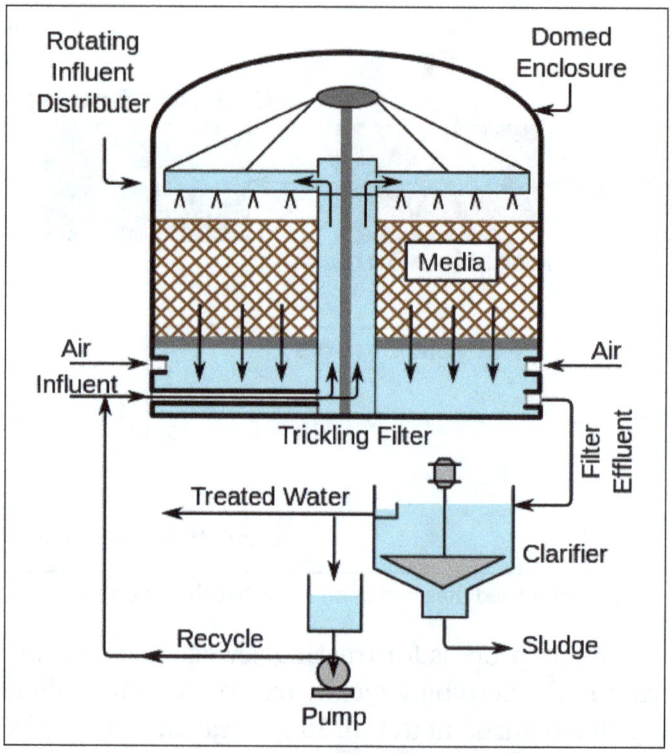

A typical complete trickling filter system.

A trickling filter consists of a bed of rocks, gravel, slag, peat moss, or plastic media over which wastewater flows downward and contacts a layer (or film) of microbial slime covering the bed media. Aerobic conditions are maintained by forced air flowing through the bed or by natural convection of air. The process involves adsorption of organic compounds in the wastewater by the microbial slime layer, diffusion of air into the slime layer to provide the oxygen required for the biochemical oxidation of the organic compounds. The end products include carbon dioxide gas, water and other products of the oxidation. As the slime layer thickens, it becomes difficult for the air to penetrate the layer and an inner anaerobic layer is formed.

The fundamental components of a complete trickling filter system are:

- A bed of filter medium upon which a layer of microbial slime is promoted and developed.
- An enclosure or a container which houses the bed of filter medium.
- A system for distributing the flow of wastewater over the filter medium.
- A system for removing and disposing of any sludge from the treated effluent.

The treatment of sewage or other wastewater with trickling filters is among the oldest and most well characterized treatment technologies.

A trickling filter is also often called a *trickle filter, trickling biofilter, biofilter, biological filter* or *biological trickling filter*.

Removal of Other Organics

Synthetic organic materials including solvents, paints, pharmaceuticals, pesticides, products from coke production and so forth can be very difficult to treat. Treatment methods are often specific to the material being treated. Methods include advanced oxidation processing, distillation, adsorption, vitrification, incineration, chemical immobilisation or landfill disposal. Some materials such as some detergents may be capable of biological degradation and in such cases, a modified form of wastewater treatment can be used.

Removal of Acids and Alkalis

Acids and alkalis can usually be neutralised under controlled conditions. Neutralisation frequently produces a precipitate that will require treatment as a solid residue that may also be toxic. In some cases, gases may be evolved requiring treatment for the gas stream. Some other forms of treatment are usually required following neutralisation.

Waste streams rich in hardness ions as from de-ionisation processes can readily lose the hardness ions in a buildup of precipitated calcium and magnesium salts. This precipitation process can cause severe *furring* of pipes and can, in extreme cases, cause the blockage of disposal pipes. A 1-metre diameter industrial marine discharge pipe serving a major chemicals complex was blocked by such salts in the 1970s. Treatment is by concentration of de-ionisation waste waters and disposal to landfill or by careful pH management of the released wastewater.

Removal of Toxic Materials

Toxic materials including many organic materials, metals (such as zinc, silver, cadmium, thallium, etc.) acids, alkalis, non-metallic elements (such as arsenic or selenium) are generally resistant to biological processes unless very dilute. Metals can often be precipitated out by changing the pH or by treatment with other chemicals. Many, however, are resistant to treatment or mitigation and may require concentration followed by landfilling or recycling. Dissolved organics can be *incinerated* within the wastewater by the advanced oxidation process.

Smart Capsules

Molecular encapsulation is a technology that has the potential to provide a system for the recyclable removal of lead and other ions from polluted sources. Nano-, micro- and milli- capsules, with sizes in the ranges 10 nm-1µm,1µm-1mm and >1mm, respectively, are particles that have an active reagent (core) surrounded by a carrier (shell).There are three types of capsule under investigation: alginate-based capsules, carbon nanotubes, polymer swelling capsules. These capsules provide a possible means for the remediation of contaminated water.

Sedimentation Wastewater Treatment

Sedimentation is the most common physical unit operation in wastewater treatment, more so in primary treatment where sedimentation is the workhorse of the treatment. The term sedimentation is also called settling in some literature. Sedimentation is, in a nutshell, a process by which the suspended solids, which have higher densities than that of water, are re-moved from wastewater by the action of gravity in the bottom of the settling tank or basin (also called a clarifier) within a reasonable period of time. Sedimentation basins are usually rectangular or circular with a radial or upward water flow pattern. Sedimentation is not limited to primary treatment; there is also secondary sedimentation by which settleable solids in the biological secondary treatment processes are removed. For example, recovery of activated sludge for recycling is achieved with secondary sedimentation.

A photo of a dissolved air flotation system.

In a typical wastewater tretment plant, the wastewater stream exiting from screening devices (and after flotation basins) then enters the second section of the primary treatment of wastewater treatment or sedimentation tanks/basins. Here, the sludge (the organic portion of the sewage) settles out of the wastewater and is pumped out of the tanks. Some of the water is removed in a step called thickening and then the sludge is processed in large tanks called digesters.

Sedimentation uses gravitational force to separate unstable and destabilized suspended solids from wastewater. It is based on the density differ ence between the bulk of the liquid and the solids. Stabilized solids such as colloids can be destabilized with flocculants. Sedimentation is a very important primary treatment process; it is, however, also used in the biological treatment, such as activated sludge and trickling filters for solid removal. The settling characteristics of the solids are determined by the types of the settling solids and their concentrations. Sedimentation has four distinct types of settling:

- Discrete settling (Type I), which is settling of a dilute suspension of solids that do not aggregate.
- Flocculent settling (Type II), which is settling of the particulates that aggregate among themselves and/or with added flocculants to form larger particulates and therefore results in faster settling. The sedimentation operation in a typical primary treatment of wastewater operates in this mode.
- Zone settling (Type III, also called hindered settling), which occurs when particulates adhere together, forming a mass that settles as a blanket with a distinguishable interface with the liquid above it.
- Compression zone (Type IV), which occurs when sinking particulates accumulate at the bottom of the sedimentation tank/basin, forming a compressed structure that supports the weight of the particulates that settled in the bottom of the tank/basin.

Although sedimentation basins in primary treatment are characterized by Type II flocculant settling, each of these zones have different characteristics that warrant further analysis.

Discrete Settling (Type I)

The settling of nonaggregated solids in a dilute suspension can be described by its settling velocity of individual particulates. In a settling tank/ basin, the settling of a discrete particle is not affected by the other particles and is only a function of the fluid property and the characteristics of the particle; this may be further depicted in Figure when the movement of the particle of interest is subject to the combined effect of the gravitational force downward and the bulk flow toward the outlet:

vt = (tank depth)/(residence time)

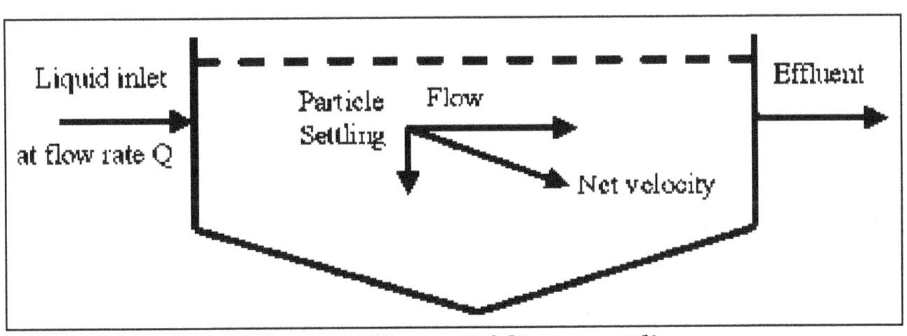

A schematic diagram of discrete settling.

or mathematically: vt = H/t

H is the depth of the sedimentation tank and t is the residence time of the particle in the liquid in the tank. If we assume the residence time of the liquid is the residence time of the liquid in the tank.

A is the cross-sectional area of the tank and Q is the overall volumetric flow rate through the tank. Here, Q/A is the overflow rate of the liquid passing through the tank.

The terminal velocity of the particle is equal to the overflow rate of the tank.

For readers who have some exposure to fluid mechanics, the above derivation might strike them as being suspicious, since the terminal velocity of a discrete particle in the liquid.

where d is the size (equivalent diameter of the particle), g is the gravitational acceleration, ^ is the viscosity of the liquid, and (ps — pl) is the density difference between the particle and the liquid.

This suggests that the terminal velocity is a function of size and the density of the particle, which is nowhere to be inferred. Note that we have assumed that the residence time of the particle in the tank is the same as that of the bulk liquid or the equivalent depth for the particle settling in the tank is the same as the depth of the tank. The overflow rate of the tank as the terminal velocity of a particle in the tank represents the critical velocity of an ideal particle in an ideal settling tank assuming:

- The number of ideal discrete particles and the velocity vectors of the liquid are uniformly distributed.
- The liquid flows in the tank as an ideal slug.
- Any particle reaching the bottom of the tank is effectively removed (no resuspension) (Canale and Borchardt, 1972).

Any ideal particles having termial velocity v (average velocity among all particles present) greater than vt is 100% removed from the settling tank/basin. For those particles with less than vt average terminal velocity, the portion of the particles removed in the tank is equal to v/vt.

In reality, the discrete settling is more likely associated with settling of hard particulates with high density and size such as grit and sand. This is a rare type of particulate in a typical food wastewater stream, but it may occur in some sources of agricultural wastewater that is subject to intrusion of soil and dirt.

Flocculent Settling (Type II)

Flocculent settling is used in primary clarifiers and the upper zones of secondary clarifiers. In the case of flocculent settling, the particles in the relatively dilute suspension coalesce or flocculate to form larger particles or aggregates during settling, thus increasing the mass of settling solids as well as the settling velocity (and removal rate). In many food wastewater treatment situations, except very dilute ones, suspended solids cannot be described as discrete particles of known specific gravity (a quantity that is the ratio of density of particle to density of water). In general, larger particles settle faster and have a greater tendency to collide with other slower-settling particles, resulting in formation of larger particles in a quiescent body of water. However, the wind, hydrodynamic shear, and hydraulic disturbance all contribute to further contacts among particles in the tank.

Sample Ports 60cm Apart

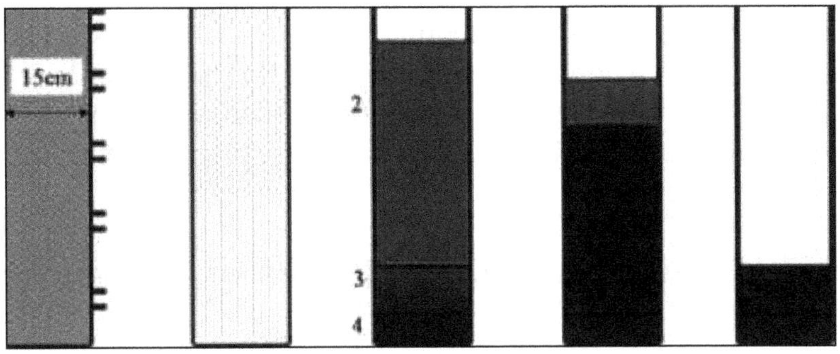

Time

A diagram of settling column and zoning settling process.

Furthermore, the greater the depth of the tank, the higher the frequency of collisons among particles will be during settling. Therefore, the flocculent settling is dependent on the properties of particles and the liquid as well as depth of the settling tank/basin. The settled solids in the bottom of the tank are usually promptly removed, so the greater rate of settling as a result of aggregation of individual particles translates into a greater rate of solid removal from the wastewater. Evaluation of a wastewater stream slated for a sedimention tank or basin is carried out using a settling column, as depicted in Figure. The laboratory of the settling column is about 15 cm (6 in) in diameter and 305 cm (10 ft) tall, and it has several sampling ports 61 cm (2 ft) apart. The settling evaluation is conducted by first placing a known quantity of wastewater sample in the column. The uniformity of particle size from top to bottom of the column in the beginning of the evaluation and the temperature of the liquid throughout the evaluation should be accomplished. The wastewater containing suspended solids is allowed to settle under quiescent conditions; small samples of suspension at different ports with preset depths are drawn and concentrations of particles in the samples are determined over preset time intervals. The frac tion removal of the particles is calculated for each sample analyzed and is plotted against time and depth. The fraction of the particles removed at each depth is constructed as curve lines called isoconcentration lines, as those illustrated in Figure. These lines represent the most efficient particle removal loci for a given removal rate. The ratio of the depth to time is the average settling (terminal) velocity of the particles under a given percent removal.

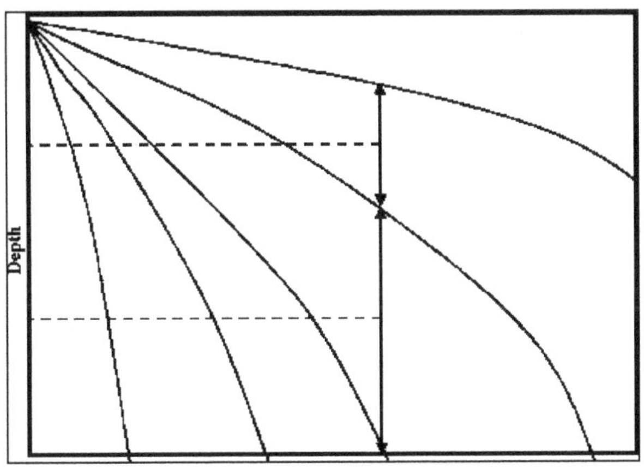

Time

A fraction of removal of flocculating particles at each depth.

Zone Settling (Type III)

Zone settling, also called hindered settling, acquires its name from the fact that aggregated particulates of a concentrated suspension (activated sludge or flocculated colloids) in the sedimentation basin tend to form a massive blanketlike suspension with a distinct interface. Zone settling is mainly used in secondary clarifiers. Many wastewater treatment process designers use a batch settling test to determine the interface.

Physicochemical Wastewater Treatment Processes 65 Compression Zone (Type IV)

Compression settling involves the highest concentration of suspended solids and occurs in the lower reaches of clarifiers. The particles settle by compressing the mass of the particles below. Compression occurs not only in the lower zones of secondary clarifiers but also in sludge thickening tanks.

Basics

Suspended solids (or SS), is the mass of dry solids retained by a filter of a given porosity related to the volume of the water sample. This includes particles 10 μm and greater.

Colloids are particles of a size between 0.001 μm and 1 μm depending on the method of quantification. Because of Brownian motion and electrostatic forces balancing the gravity, they are not likely to settle naturally.

The limit sedimentation velocity of a particle is its theoretical descending speed in clear and still water. In settling process theory, a particle will settle only if :-

1. In a vertical ascending flow, the ascending water velocity is lower than the limit sedimentation velocity.
2. In a longitudinal flow, the ratio of the length of the tank to the height of the tank is higher than the ratio of the water velocity to the limit sedimentation velocity.

Removal of suspended particles by sedimentation depends upon the size and specific gravity of those particles. Suspended solids retained on a filter may remain in suspension if their specific gravity is similar to water while very dense particles passing through the filter may settle. Settleable solids are measured as the visible volume accumulated at the bottom of an Imhoff cone after water has settled for one hour.

Gravitational theory is employed, alongside the derivation from Newton's second law and the Navier–Stokes equations.

Stokes' law explains the relationship between the settling rate and the particle diameter. Under specific conditions, the particle settling rate is directly proportional to the square of particle diameter and inversely proportional to liquid viscosity.

The settling velocity, defined as the residence time taken for the particles to settle in the tank, enables the calculation of tank volume. Precise design and operation of a sedimentation tank is of high importance in order to keep the amount of sediment entering the diversion system to a minimum threshold by maintaining the transport system and stream stability to remove the sediment diverted from the system. This is achieved by reducing stream velocity as low as possible for the longest period of time possible. This is feasible by widening the approach channel and lowering its floor to reduce flow velocity thus allowing sediment to settle out of suspension due to gravity. The settling behavior of heavier particulates is also affected by the turbulence.

Designs

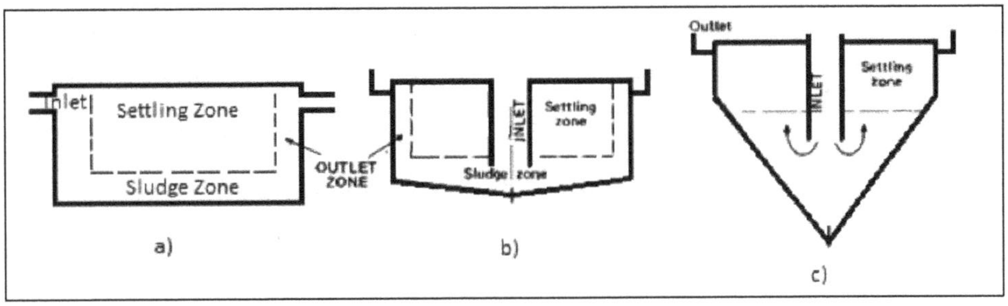

Different clarifier designs.

Although sedimentation might occur in tanks of other shapes, removal of accumulated solids is easiest with conveyor belts in rectangular tanks or with scrapers rotating around the central axis of circular tanks. Settling basins and clarifiers should be designed based on the settling velocity of the smallest particle to be theoretically 100% removed. The overflow rate is defined as:

Overflow rate (V_o) = Flow of water (Q (cubic metre per second)) /(Surface area of settling basin (A))(m^2)

In many countries this value is named as surface loading in m3/h per m2. Overflow rate is often used for flow over an edge (for example a weir) in the unit m3/h per m.

The unit of overflow rate is usually meters (or feet) per second, a velocity. Any particle with settling velocity (*Vs*) greater than the overflow rate will settle out, while other particles will settle in the ratio Vs/Vo. There are recommendations on the overflow rates for each design that ideally take into account the change in particle size as the solids move through the operation:

- Quiescent zones: 9.4 mm (0.031 ft) per second,
- Full-flow basins: 4.0 mm (0.013 ft) per second,
- Off-line basins: 0.46 mm (0.0015 ft) per second.

However, factors such as flow surges, wind shear, scour, and turbulence reduce the effectiveness of settling. To compensate for these less than ideal conditions, it is recommended doubling the area calculated by the previous equation. It is also important to equalize flow distribution at each point across the cross-section of the basin. Poor inlet and outlet designs can produce extremely poor flow characteristics for sedimentation.

Settling basins and clarifiers can be designed as long rectangles (Figure. a), that are hydraulically more stable and easier to control for large volumes. Circular clarifiers (Figure. b) work as a common thickener (without the usage of rakes), or as upflow tanks (Figure. c).

Sedimentation efficiency does not depend on the tank depth. If the forward velocity is low enough so that the settled material does not re-suspend from the tank floor, the area is still the main parameter when designing a settling basin or clarifier, taking care that the depth is not too low.

Assessment of Main Process Characteristics

Settling basins and clarifiers are designed to retain water so that suspended solids can settle. By sedimentation principles, the suitable treatment technologies should be chosen depending on the specific gravity, size and shear resistance of particles. Depending on the size and density of particles, and physical properties of the solids, there are four types of sedimentation processes:

- Type 1 – Dilutes, non-flocculent, free-settling (every particle settles independently).
- Type 2 – Dilute, flocculent (particles can flocculate as they settle).
- Type 3 – Concentrated suspensions, zone settling, hindered settling (sludge thickening).
- Type 4 – Concentrated suspensions, compression (sludge thickening).

Different factors control the sedimentation rate in each.

Settling of Discrete Particles

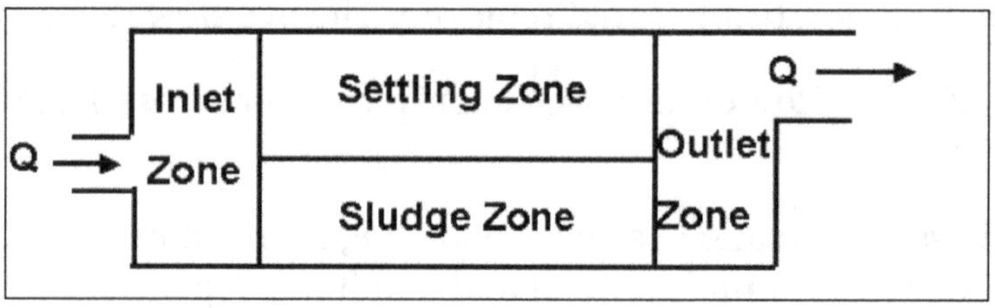

The four functional zones of a continuous flow settling basin.

Unhindered settling is a process that removes the discrete particles in a very low concentration without interference from nearby particles. In general, if the concentration of the solutions is lower than 500 mg/L total suspended solids, sedimentation will be considered discrete. Concentrations of raceway effluent total suspended solids (TSS) in the west are usually less than 5 mg/L net. TSS concentrations of off-line settling basin effluent are less than 100 mg/L net. The particles keep their size and shape during discrete settling, with an independent velocity. With such low concentrations of suspended particles, the probability of particle collisions is very low and consequently the rate of floculation is small enough to be neglected for most calculations. Thus the surface area of the settling basin becomes the main factor of sedimentation rate. All continuous flow settling basins are divided into four parts: inlet zone, settling zone, sludge zone and outlet zone.

In the inlet zone, flow is established in a same forward direction. Sedimentation occurs in the settling zone as the water flow towards to outlet zone. The clarified liquid is then flow out from outlet zone. Sludge zone: settled will be collected here and usually we assume that it is removed from water flow once the particles arrives the sludge zone.

In an ideal rectangular sedimentation tank, in the settling zone, the critical particle enters at the top of the settling zone, and the settle velocity would be the smallest value to reach the sludge zone, and at the end of outlet zone, the velocity component of this critical particle are Vs, the settling velocity in vertical direction and Vh in horizontal direction.

From Figure, the time needed for the particle to settle;

$$t_o = H/V_p = L/Vs \quad (3)$$

Since the surface area of the tank is WL, and Vs = Q/WL, Vh = Q/WH, where Q is the flow rate and W, L, H is the width, length, depth of the tank.

According to Equation, this also is a basic factor that can control the sedimentation tank performance which called overflow rate.

Equation also tell us that the depth of sedimentation tank is independent to the sedimentation efficiency, only if the forward velocity is low enough to make sure the settled mass would not suspended again from the tank floor.

Settlement of Flocculent Particles

In a horizontal sedimentation tank, some particles may not follow the diagonal line in Figure, while settling faster as they grow. So this says that particles can grow and develop a higher settling velocity if a greater depth with longer retention time. However, the collision chance would be even greater if the same retention time were spread over a longer, shallower tank. In fact, in order to avoid hydraulic short-circuiting, tanks usually are made 3–6 m deep with retention times of a few hours.

Zone-settling Behaviour

As the concentration of particles in a suspension is increased, a point is reached where particles are so close together that they no longer settle independently of one another and the velocity fields of the fluid displaced by adjacent particles, overlap. There is also a net upward flow of liquid displaced by the settling particles. This results in a reduced particle-settling velocity and the effect is known as hindered settling.

There is a common case for hindered settling occurs. the whole suspension tends to settle as a 'blanket' due to its extremely high particle concentration. This is known as zone settling, because it is easy to make a distinction between several different zones which separated by concentration discontinuities. Figure represents a typical batch-settling column tests on a suspension exhibiting zone-settling characteristics. There is a clear interface near the top of the column would be formed to separating the settling sludge mass from the clarified supernatant as long as leaving such a suspension to stand in a settling column. As the suspension settles, this interface will move down at the same speed. At the same time, there is an interface near the bottom between that settled suspension and the suspended blanket. After settling of suspension is complete, the bottom interface would move upwards and meet the top interface which moves downwards.

Compression Settling

The settling particles can contact each other and arise when approaching the floor of the sedimentation tanks at very high particle concentration. So that further settling will only occur in adjust matrix as the sedimentation rate decreasing. This is can be illustrated by the lower region of the zone-settling diagram. In Compression zone, the settled solids are compressed by gravity (the weight of solids), as the settled solids are compressed under the weight of overlying solids, and water is squeezed out while the space gets smaller.

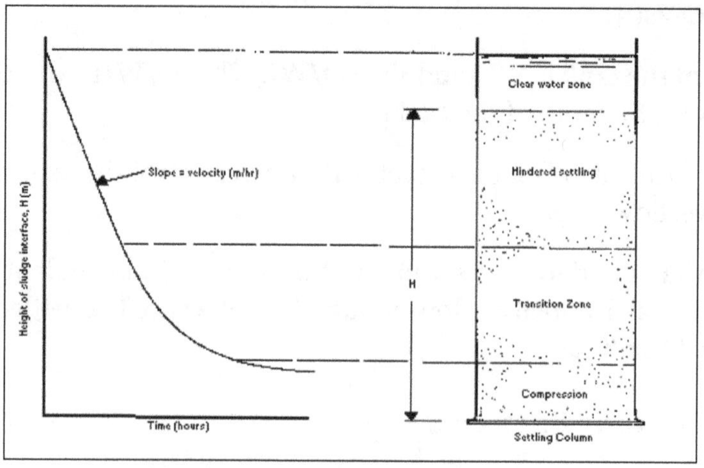

Typical batch-settling column test on a suspension exhibiting zone-settling characteristics.

Sedimentation Tank

Sedimentation tank, also called settling tank or clarifier, component of a modern system of water supply or wastewater treatment. A sedimentation tank allows suspended particles to settle out of water or wastewater as it flows slowly through the tank, thereby providing some degree of purification. A layer of accumulated solids, called sludge, forms at the bottom of the tank and is periodically removed. In drinking-water treatment, coagulants are added to the water prior to sedimentation in order to facilitate the settling process, which is followed by filtration and other treatment steps. In modern sewage treatment, primary sedimentation must be followed by secondary treatment (e.g., trickling filter or activated sludge) to increase purification efficiencies. Sedimentation is usually preceded by treatment using bar screens and grit chambers to remove large objects and coarse solids.

Different Types of Sedimentation

There are various different types of sedimentation:

- Granular particles will each settle separately and will each be subject to a constant settling speed;
- Particles that are more or less flocculated will be of different sizes and, therefore, subject to variable sedimentation rates. At low concentration levels, the settling speed increases as the floc increases in size through collision with other particles; this is termed flocculent settling.

At higher concentration levels, the abundance of floc and its interactions will lead to an overall sedimentation which is most frequently characterised by a clearly defined interface between the sludge mass and the supernatant liquid: this is hindered settling where the rate will be optimum within a certain area of concentration, above which, we talk of hindered settling.

Granular Particle Sedimentation

This is the simplest case and the only one that can be easily described by equations.

Theory: Fluid at Rest

When a granular particle remains in a liquid at rest, it will be affected by a driving force F_M (gravity minus the Archimedes thrust) and a resisting force F_T (fluid drag) created by viscosity and inertia forces:

$$F_M = g \cdot v \cdot \Delta\rho$$

$$F_T = \frac{C \cdot s \cdot \rho_F \cdot V^2}{2}$$

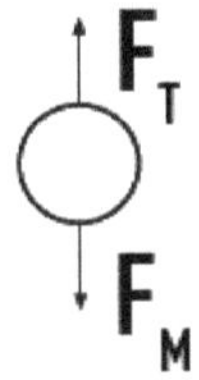

ρ_p, ρ_L: density of the granular particle and of the fluid, and $\Delta\rho = \rho_p - \rho_L$,

d,s,v : diameter, projected surface area (master couple: $\frac{\pi d^2}{4}$ for a sphere)

and volume of the granular particle.

V: particle settling rate,
g: gravity acceleration,
C: drag coefficient (adimensional),
Very quickly, a ($F_M = F_T$) balance will become established and the sedimentation of the particle treated like a sphere will occur at a constant rate V_I such that:

$$V_0^2 = \frac{4}{3} g \frac{d}{C} \frac{\Delta\rho}{\rho_F}$$

Hydraulic Conditions

The value of C, the drag coefficient, is defined by the disturbance which is in turn based on the settling speed. This disturbance is characterised by the Reynolds grain number established by:

$$Re = \frac{\rho_F \cdot V \cdot d}{\mu}$$

Re = adimensional.
where μ = dynamic viscosity.

When Re is low, the viscosity forces are far greater than the inertia forces. When Re is high, the viscosity forces are negligible.

The drag coefficient is provided by:

$$C = a \cdot Re^{-n}$$

Table provides the various values for a, n and C according to the Reynolds number.

Re	Condition	a	n	C	Formula
$10^4 < Re < 1$	Laminar	24	1	$24 \cdot Re^{-1}$	Stokes
$1 < Re < 10^3$	Intermediate	18.5	0.6	$18.5 \cdot Re^{-0.6}$	Allen
$10^3 < Re < 2 \cdot 10^5$	Turbulent	0.44	0	0.44	Newton

Table 9. Hydraulic conditions

These formulas constitute the basis used for calculating grain movement in a fluid and are used for sedimentation (granular solids in a liquid, water drops in air), in upward motion (air bubbles in water, oil drops in water), for centrifugation, and for fluidisation.

Under laminar conditions, the Stokes law applied to a spherical particle will give:

$$V_0 = \frac{g}{18 \cdot \mu} \Delta\rho \cdot d^2$$

The aggregation phenomena that cause growth will, therefore, very rapidly increase the sedimentation rate.

Under transitory conditions, the Allen law also gives a rising rate based on particle size; however, this rate rises far more slowly because:

$$V_0^{1.4} = \frac{g}{13.875 \cdot \mu^{0.6}} \frac{\Delta\rho}{\rho_F^{0.4}} d^{16}$$

$$\text{then } V_0 = k \cdot d^{1,143}$$

Sphericity Factor

This factor Ψ is provided by:

$$\Psi = \frac{\text{volume of the sphere having the same surface}}{\text{grain volume}}$$

In the preceding operations, we then need to replace C by C' = ΨC and the Stokes law will be written:

$$V_0 = \frac{g}{18 \cdot \mu} \Delta\rho \cdot d^2$$

Table illustrates the considerable effect this factor has on "flat" materials.

Values for ψ	
Sand	2
Carbon	2.25
Talc	[illegible]
Gypsum	4
Graphite lamella	22
Mica	170

The considerable effect this factor has on "flat" materials.

Recovery Conditions

Let us consider a rectangular sedimentation tank that has a length L, a vertical section $S = H \cdot \ell$ (where H is the water depth and $\ell$the width) and a horizontal section $S_H = L \cdot \ell$, evenly crossed by a throughput Q that either ascends vertically or travels horizontally; the following conditions will apply if the tank is to screen out a granular particle settling at a rate V_o in still water:

Vertical Upflow Sedimentation

Particles having a sedimentation rate that is greater than the liquid's upwards velocity will be screened out. This phenomenon is written:

$$V_0 > V_{asc} = \frac{Q}{S_H}$$

Q = liquid flow rate.
S_H = settling tank free surface area.

horizontal flow sedimentation

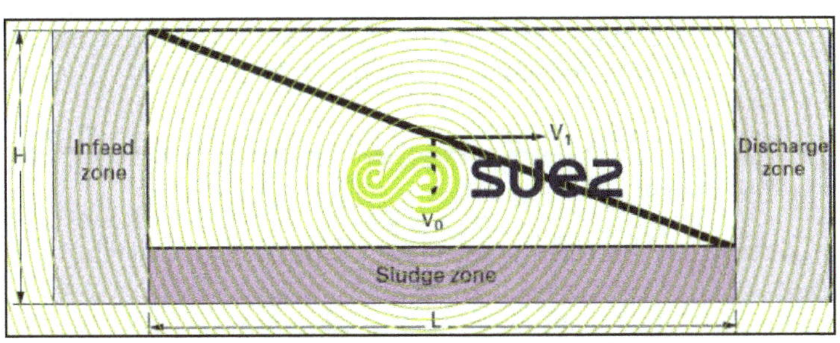

Schematic diagram of a horizontal flow sedimentation application (granular particles).

The velocity of a particle entering into the tank through the top will have two components:

$V_{1:}$ fluid horizontal velocity equal to Q/S.

$V_{0:}$ vertical settling speed provided by Stokes law.

This particle will be retained in the tank when the time it takes for the particle to touch the tank floor (or to penetrate into the sludge zone)

$$t_1 = \frac{H}{V_0} \text{ is less than the water's contact time in the sedimentation tank } t_2 = \frac{L}{V_1} = \frac{L \cdot S}{Q}.$$

Then:

$$\frac{H}{V_0} < \frac{L \cdot S}{Q} \text{ or } V_0 > \frac{H \cdot Q}{L \cdot S} = \frac{H \cdot Q}{L \cdot \ell \cdot H} = \frac{Q}{S_H} = V_H$$

$V_{H:}$Hazen velocity (or hydraulic loading on the surface) similar to Vasc in the preceding example and expressed in $m^3 \cdot (h{\cdot}m^2)^{-1}$ or $m \cdot h^{-1}$.

It should be noted that V_H is independent of tank depth.

All particles having sedimentation rates above V_H will, theoretically, be eliminated. However, if the water infeed is distributed over the entire depth of the tank, some of the particles that have a sedimentation rate V below the Hazen velocity will also be retained within the V/V_H ratio whereas these particles would not have been retained in a upward flow sedimentation tank.

In theory, for identical horizontal surface areas, a horizontal flow sedimentation tank can thus be used to separate a greater number of particles.

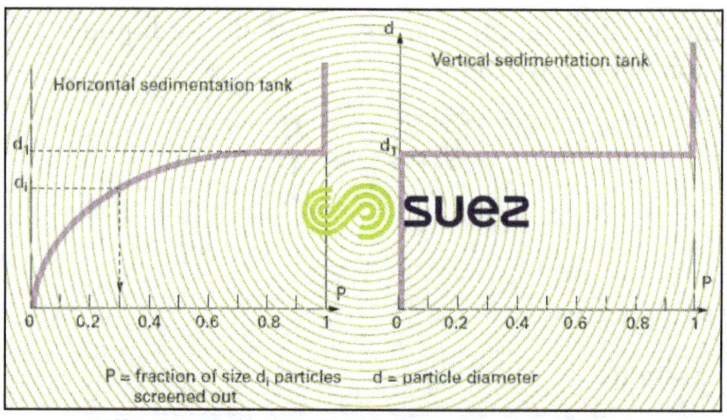

Effectiveness comparison between horizontal and upward flow sedimentation (granular particles).

In practice, this difference will be attenuated and even reversed for the following reasons associated with horizontal flow sedimentation:

- Hydraulic distribution problems in the vertical plane both at the inlet and at the outlet of a structure;
- Sludge accumulation and collection, reducing the available section;

- In a circular, horizontal flow sedimentation tank, the horizontal component of the particle velocity (V_1) decreases from the centre outwards and the particle will adopt a curvilinear trajectory.

Flocculent Settling of Flocculated Particles

During sedimentation, flocculation will continue to take place and particle sedimentation rate V_o will rise.

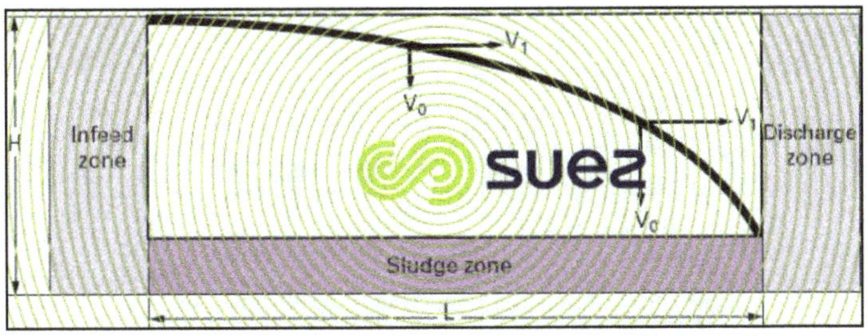

Schematic diagram of a horizontal flow sedimentation application (flocculated particles).

This process takes place as soon as flocculated matter concentration rises above approximately 50 $mg \cdot L^{-1}$.

Flocculent sedimentation effectiveness does not depend only on the hydraulic loading at the surface but also on contact time. There are no mathematical formulae available for calculating the sedimentation rate.

Only laboratory tests and graph methods can be used to ascertain this rate. Figure provides the results of one such test.

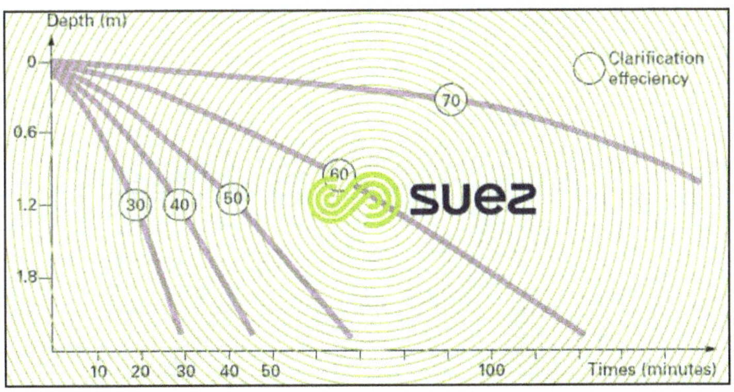

Eliminating flocculated particles by the flocculent sedimentation method: relationship between time, effective depth and sedimentation performance.

Hindered Settling of Flocculated Particles

When the flocculated particle concentration increases, the interaction between particles can no longer be ignored; they undergo "hindered" settling. Initially, this may cause flocculation and sedimentation to improve and then hindered beyond a certain critical concentration; we then talk of "hindered settling".

This is a phenomenon that is characteristic of activated sludge and flocculated suspensions when their concentration rises above approximately 500 mg · L^{-1}.

Visual Observation

When hindered settling is carried out in a tube of adequate height and diameter (at least a 1-litre test tube), we normally see the appearance of four zones.

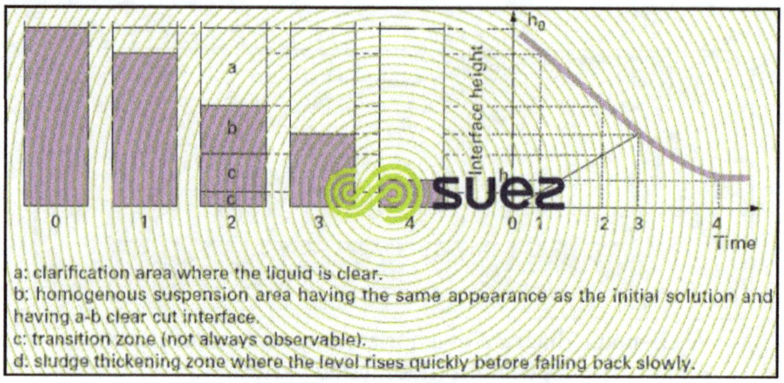

Hindered settling: Kynch curve.

After a certain state, zones b and c disappear; this is the critical point. The change in the a-b interface height as a function of time constitutes the Kynch curve.

Kynch Curve

Kynch's fundamental hypothesis is that the rate at which a particle falls depends exclusively on local particle concentration C Courbe de Kynch.

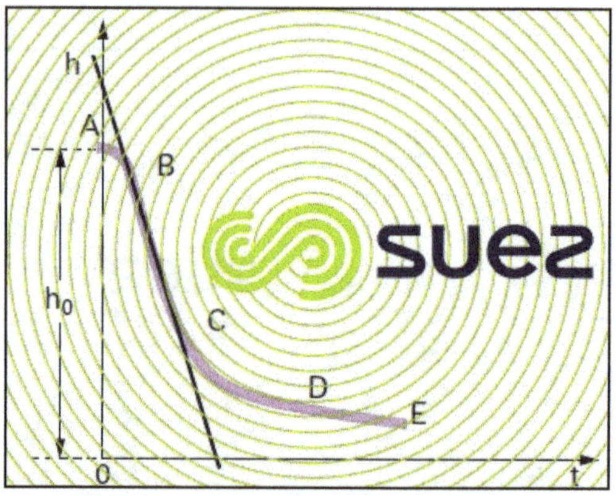

Kynch curve.

The separation surface between A and B is more or less clear-cut: this is the flake coalescence phase. This phase does not always exist.

From B to C, a rectilinear part identifies a constant settling speed V (straight line slope). In the case of a tube of specific dimensions, V depends on the initial suspended solids concentration and on the suspension's flocculation properties. When the initial concentration C increases, the mass's sedimentation rate V decreases: e.g. in activated urban sludge where the suspended solids concentration goes from 1 to 4 g · L^{-1}, Vdecreases from 6 to 1.8 m · h^{-1}.

The CD section which is concave at the top, refers to a gradual decrease in the settling speed of the deposit's top layer.

From D, the flakes come into contact with each other and exert a compression effect on the lower layers.

Kynch's theory applies to section BC and CD that cover the main flocculated sludge sedimentation area.

Interpretation

Let us consider a suspension where its clarification does not include a coalescence phase, calculations show that:

- In the BOC triangle, concentration and settling speed remain constant and equal to the initial values found in B;
- In the COD triangle, the equal concentration curves are straight lines that cross through the origin; this means that, as soon as sedimentation commences, the layers that are the closest to the base of the tank will have to transit through every concentration from initial concentration to that applicable at point D, the start of compression.

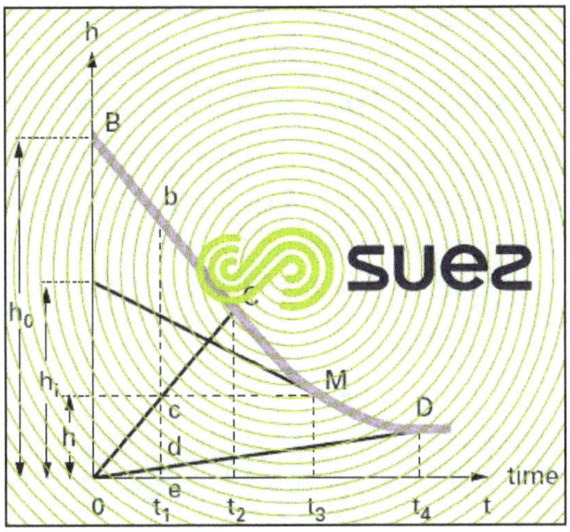

Interpretation.

Consequently, the sludge medium that has a height eb at point in time t_1 will have three separate zones:

- An upper zone bc where settling speed and concentration remain uniform and will have retained their initial values V_0 and C;
- An intermediate zone cd where concentration increases gradually from c to d and the settling speed drops accordingly;
- A lower zone where sludge flakes are subject to compression.

In the medium considered at a point in time t_2, the upper zone disappears and, at point in time t_4, only the lower zone remains.

With regard to point M in section CD, there are two concentration levels:

C_M^i concentration at the interface,

C_M mean concentration.

According to the Kynch hypothesis:

$$C_M^I = C_o \frac{h_o}{h_i}$$

Furthermore:

$$C_M^I = C_o \frac{h_o}{h}$$

The three sections BC, CD and DE on the Kynch curve (figure 16) are used to dimension hindered settling tanks. The BC phase refers to the solids contact clarification area. The CD phase applies to structures where sludge concentration is targeted (units used for thickened sludge recirculation). The DE phase is used for sludge thickening.

Mohlman Index (Sludge Volume Index: Svi)

One specific point needs to be considered on the Kynch curve, that of the 30 minute abscissa: the Mohlman index I_M is widely used to define that clarification capacity of biological sludge and, therefore, for dimensioning its clarifier and even for initiating remedial action should bulking occur, see section biomass used in wastewater purification and chapter biological processes).

$$I_M = \frac{V}{M} \ (cm^3 \cdot g^{-1})$$

V: volume of sludge after 30 minute settling time (cm^3),
M: suspended solids present in this volume (g).

The drawback of the Mohlman index is that it is heavily dependent on sludge initial concentration. Therefore, methods have been put forward for establishing an independent index for this concentration and which is, therefore, exclusively characteristic of the state of the sludge in the plant concerned. This is:

Sludge Index (IB) or Diluted Mohlman Index (DSVI)

The Kynch curve is plotted from dilutions used to achieve a volume of sludge equal to approximately 200-250 mL per L after 30 minutes of settling time. The sludge index I_B used in France applies when this volume is between 100 and 300 mL; the DSVI used by the Anglo-Saxons applies when the volume is equal to between 150 and 250 mL. In both cases, it can be said that activated sludge has very good settling properties when its sludge index or its DSVI is between 50 and 100 $cm^3 \cdot g^{-1}$, normal settling properties between 100 and 200 and poor settling properties above 250.

Designing Settling Tanks

Two criteria are used for the purpose of calculating the surface area of a settling tank:

- The hydraulic surface loading that is characteristic of the volume of water that can be processed by the unit of surface area and unit of time ($m^3 \cdot m^{-2}.h^{-1}$);
- The solids loading that is characteristic of suspended solids that will settle per unit of surface area and per unit of time ($kg \cdot m^{-2}.h^{-1}$).

The most limiting of the two calculations will constitute the structures dimensioning parameter.

Influence of the Hydraulic Surface Loading

This loading is directly linked to "free or flocculated" suspended solids sedimentation rate and the preceding paragraph"different types of sedimentation" shows how this rate can be determined and how to calculate the ensuing minimum surface area.

Influence of Solids Loading

In flocculated particle hindered settling which includes thickening phenomena, mass flow rate is usually a determining factor when calculating settling tank surface area.

Let us consider a settling tank having a section S, fed with an incoming flow Q_E with a suspended solids concentration C_E; sludge is drawn off from the bottom of the settling tank at a flow rate of Q_S with a concentration C_S.

In the absence of chemical or biological reactions affecting suspended solids concentration, and if we consider a 100% elimination efficiency, we will have:

- Treated flow $Q = Q_E - Q_S$
- Materials balance $\cdot C_S = Q_E \cdot C_E$

or as solids loading:

$$\frac{Q_S \cdot C_S}{S} = \frac{Q_E \cdot C_E}{S}$$

The solids loading that will settle is provided by the Kynch curve. For a given point on the Kynch curve, having a concentration C_i, the sedimentation rate V_i is provided by the tangent at this point. The corresponding mass flow is $F_i = C_i \cdot V_i$.

To this mass flow F_i, we need to add the draw off mass flow F_S provided by $C_iV_s = Q_s/S$.

The total solids loading will be $F = C_i{\cdot}V_i + C_i{\cdot}V_s$.

Figure illustrates the changes undergone by these different flows. Flow F has a minimum F_L which is associated with a critical concentration C_L, which requires the settling tank to have a minimum section of S_m such that:

$$S_m = \frac{Q_E \cdot C_E}{F_L}$$

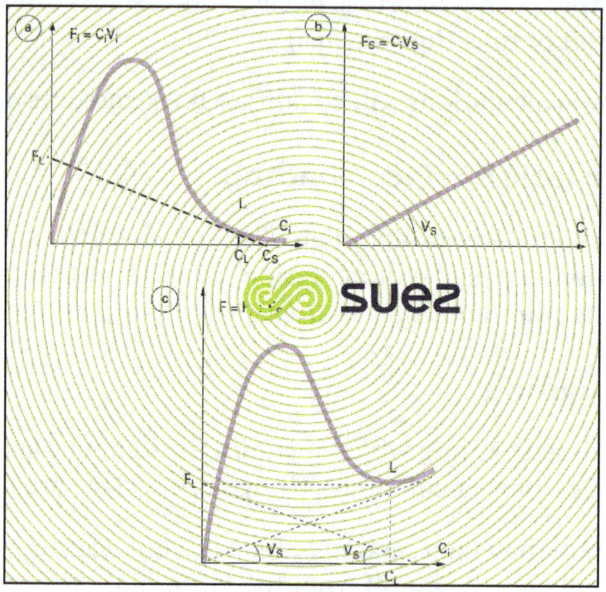

Solids loading curves.

This specific point L can be determined direct on the solids loading curve F (figure 18 c) using:

$$\left(\frac{dF}{dC_i}\right)_L = \left(\frac{dF_i}{dC_i}\right)_L + V_S = 0$$

Point L is, therefore, the point on flow curve Fi where the slope of the tangent is equal in absolute value to the draw-off speed V_s. These results can be expressed in a variety of ways by considering the Kynch curve. Limit flow F_L at point L is provided by:

$$F_L = C_L(V_L + V_S) = C_L\left(V_L + \frac{Q_E \cdot C_E}{S} \times \frac{1}{C_S}\right)$$

V_L being the sedimentation rate at point L.

Therefore, for settling to be feasible:

$$\frac{Q_E \cdot C_E}{S} < \frac{V_L}{\frac{1}{C_L} - \frac{1}{C_S}}$$

Settling Tank Structure

$$R_{e^*} = \frac{V \cdot d_h}{\upsilon}$$

Re* : Reynolds number,
V : water circulation velocity ($m \cdot s^{-1}$),
d_h : equivalent hydraulic diameter (m),
υ : water's kinematic viscosity ($m^2 \cdot s^{-1}$),
with:

$$d_h = 4 \frac{\text{wet surface}}{\text{wet perimeter}}$$

In practice, the ideal settling tank does not exist: turbulence will occur within the liquid, especially in the intake zone and wind can create waves on the surface of the liquid; convection currents created by local temperatures (sunshine) and density differentials will have an effect on clarification performance. Wherever possible, the aim must be to obtain a stable, laminar circulation characterised by an appropriate Reynolds number values established by:

Note: In the case of a full circular pipe, the hydraulic diameter will be the pipe's diameter.

In practice, conditions are deemed to be laminar when $Re^* < 800$.

Furthermore, the Froude number can be used to evaluate circulation process stability when the main influence on the flow is generated by gravity and inertia forces.

$$Fr = \frac{V^2}{g \cdot d_h}$$

The more stable the circulation, the more even the velocity distribution over the entire tank section and the better the separation. However, stable circulation can be characterised by high Froude numbers.

In practice, we can establish optimum H/L or H/R ratios where H is the wetted depth of rectangular settling tanks having a length L or of circular settling tanks having a radius R. Taking a 2-hour contact time, Schmidt-Bregas makes the following recommendations:

- Horizontal flow, rectangular settling tanks:

$$\frac{1}{35} < \frac{H}{L} < \frac{1}{20}$$

- Circular settling tanks:

$$\frac{1}{8} < \frac{H}{R} < \frac{1}{6}$$

The shape of the structure, the design of the mechanism for supplying raw water and of the mechanism for collecting treated water, as well as the sludge removal mode will all have a significant impact on settling tank performance.

In the case of water or liquor that is heavily loaded with suspended solids, the "density currents" can produce velocity distributions that tend to cause suspended solids that have accumulated on the tank floor to rise in the direction of the recovery channel. This is, for instance, the case of conventional rectangular or circular settling tanks that are too long and that are used to clarify activated sludge.

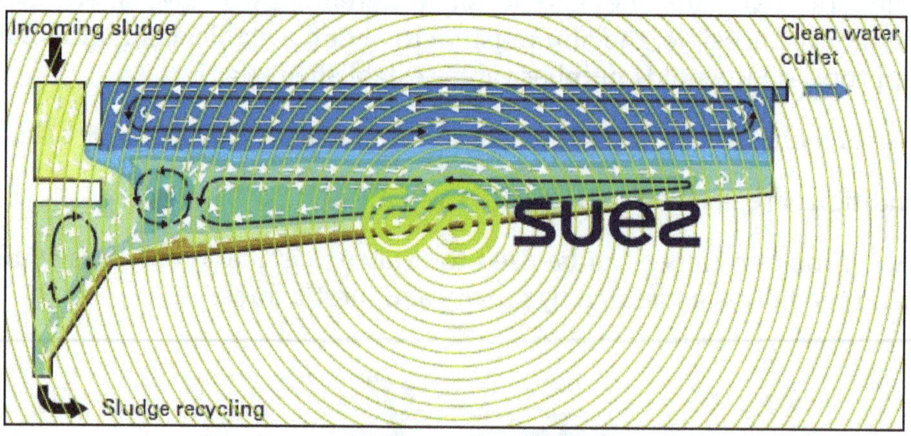

Density currents in a settling tank (CFD study).

Similarly, the effects of temperature (sunshine, water subject to rapid temperature fluctuations) and disturbances linked to changes in salinity (estuary water, IWW) will inject variable density water into the settling tank, creating convection currents, even completely overturning the water mass in the settling tank.

Clarifier

Clarifiers are required where ever the Suspended solids in raw/waste water are higher in concentration. Almost all treatment plant (Clarifiers) sedimentation tanks of circular or sometimes rectangular design.

Clarifiers work on the principle of gravity settling. The heavier suspended solids settle in the clarifier due to the quiescent conditions provided in the Clarification zone. The settled solids are swept to the centre well provided for collection of sludge with help of moving scraper blades.

Many a times the natural settling is enhanced by addition of coagulant & polyelectrolyte. The coagulant neutralizes the charges & agglomerates the suspended solids to form micro-floc. The polyelectrolyte brings together these micro-flocs binding them with long chains to make heavy floc which easily settles down. Most of the waste waters contain some scum material which does not settle down & needs to be collected on the surface of the clarifier. Hence a scum removal arrangement is normally provided. A simple clarifier is depicted in sketch below.

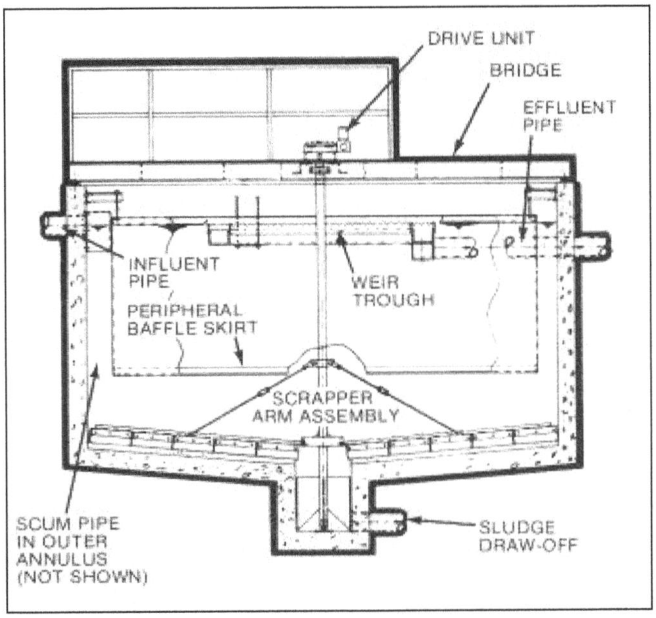

To facilitate the dosing of Chemicals as well as to ensure proper flocculation, many clarifiers are equipped with either separate flocculator or some clarifiers have a flocculating zone also.

Configurations

The clarifiers are designed based on the intended function & the space availability. In conventional waste water treatment plants there are normally two types:

a. Primary Clarifier.

b. Secondary Clarifier.

Specific applications require the following:

a. Lamella Clarifier.

b. Solids Contact Clarifier.

There are many design variation among the above type but available space & flow dictates whether it will be a circular or rectangular clarifier.

Circular Clarifier

The Clarifier design can be applied to water or wastewater treatment systems. It includes a larger influent well to provide the required flocculation time. Mixing is also added to achieve economical flocculation.

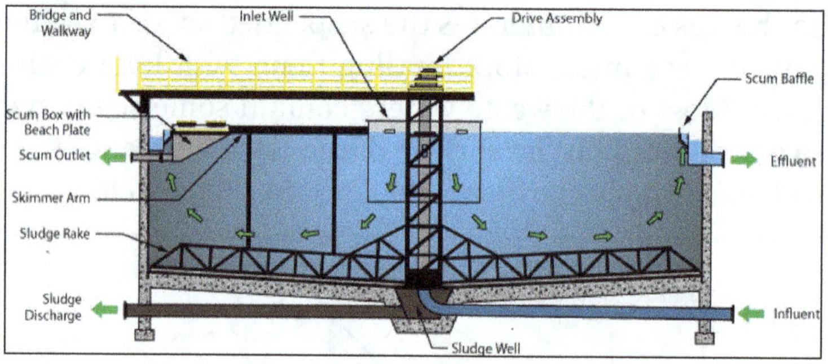

Designed to exacting specifications, the Clarifier is shop fabricated for bolted assembly, comprises of a reinforced influent well and features cast iron drive housings, deep scraper blades, adjustable squeegees and full surface adjustable skimmers.

Mechanical flocculation is proved by either concentric 'stacked' drives or independent mixers. Tanks start at ten feet in diameter and larger. Circular sludge collectors are available in either full or half bridge designs.

Rectangular Clarifier

The Chain & Scraper Sludge Collection System provides maximum sludge concentrations and scum/floating solid removal regardless of the size or application.

Rectangular clarification, a separation process commonly used in very large or space constrained municipal and industrial spaces, removes both settled and suspended solids from liquids.

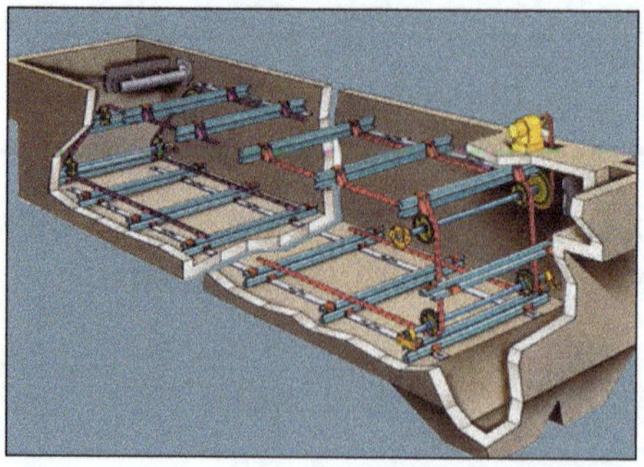

Suitable for primary, secondary, storm water collection and water plant service, Chain & Scraper is a high-quality option for both new and existing rectangular clarifier installations. To operate, flights mounted on two parallel strands of metallic or non-metallic chain scrape the settled solids along the tank floor to sludge hoppers. On the return run, the flights can skim the surface and concentrate the floating material at a scum removal device. Both three and two shaft designs are available if skimming is not a necessary function.

To assist the gravity settling many innovations are made in clarifier design.

A very popular design has Inclined plates (Lamella Clarifier) in the clarifying zone for rapid settling of flocs. The plates are placed at an angle (of 55-65°) & the water travels upwards. The solids (flocs) due to inherent weight cannot travel with same velocity & tend to settle on the plates provided. The slope of the plates ensures that the solids slide down to the collection chamber.

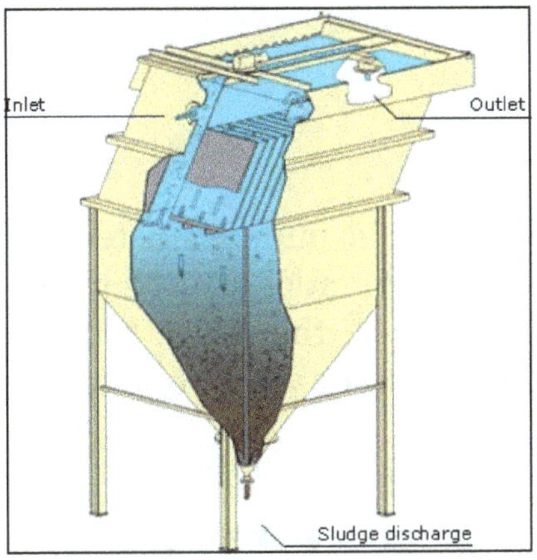

A schematic of Lamella Clarifier.

Another design involves creating slow turbulence in the settled sludge itself. The upward moving sludge helps to bring down the settleable flocs faster to sludge zone.

Applications

Pretreatment

Before the water enters the clarifier, coagulation and flocculation reagents, such as polyelectrolytes and ferric sulfate, can be added. These reagents cause finely suspended particles to clump together and form larger and denser particles, called flocs, that settle more quickly and stably. This allows the separation of the solids in the clarifier to occur more efficiently and easily; aiding in the conservation of energy. Isolating the particle components first using these processes may reduce the volume of downstream water treatment processes like filtration.

Potable water treatment

Water being purified for human consumption, is treated with floculation reagents, then sent to the clarifier where removal of the flocculated coagulate occurs producing clarified water. The clarifier works by permitting the heavier and larger particles to settle to the bottom of the clarifier. The particles then form a bottom layer of sludge requiring regular removal and disposal. Clarified water then proceeds through several more steps before being sent for storage and use.

Waste water treatment

Sedimentation tanks have been used to treat wastewater for millennia.

Primary treatment of sewage is removal of floating and settleable solids through sedimentation. *Primary clarifiers* reduce the content of suspended solids and pollutants embedded in those suspended solids. Because of the large amount of reagent necessary to treat domestic wastewater, preliminary chemical coagulation and flocculation are generally not used, remaining suspended solids being reduced by following stages of the system. However, coagulation and flocculation can be used for building a compact treatment plant (also called a "package treatment plant"), or for further polishing of the treated water.

Sedimentation tanks called *secondary clarifiers* remove flocs of biological growth created in some methods of secondary treatment including activated sludge, trickling filters and rotating biological contactors.

Mining

Methods used to treat suspended solids in mining wastewater include sedimentation and floc blanket clarification and filtration. Sedimentation is used by Rio Tinto Minerals to refine raw ore into refined borates. After dissolving the ore, the saturated borate solution is pumped into a large settling tank. Borates float on top of the liquor while rock and clay settles to the bottom.

Technology

Rectangular sedimentation tanks with effluent weir structure visible above the fluid surface.

Drained circular sedimentation tank showing central inlet baffles on the right with solids scraper and skimmer arms visible under the rotating bridge.

Although sedimentation might occur in tanks of other shapes, removal of accumulated solids is easiest with conveyor belts in rectangular tanks or with scrapers rotating around the central axis of circular tanks. Mechanical solids removal devices move as slowly as practical to minimize resuspension of settled solids. Tanks are sized to give water an optimal residence time within the tank. Economy favors using small tanks; but if flow rate through the tank is too high, most particles will not have sufficient time to settle, and will be carried with the treated water. Considerable attention is focused on reducing water inlet and outlet velocities to minimize turbulence and promote effective settling throughout available tank volume. Baffles are used to prevent fluid velocities at the tank entrance from extending into the tank; and overflow weirs are used to uniformly distribute flow from liquid leaving the tank over a wide area of the surface to minimize resuspension of settling particles.

Tube Settlers

Tube settlers are commonly used in rectangular clarifiers to increase the settling capacity by reducing the vertical distance a suspended particle must travel. High efficiency tube settlers use a stack of parallel tubes, rectangles or flat pieces separated by a few inches (several centimeters) and sloping upwards in the direction of flow. This structure creates a large number of narrow parallel flow pathways encouraging uniform laminar flow as modeled by Stokes' law. These structures work in two ways:

1. They provide a very large surface area onto which particles may fall and become stabilized.
2. Because flow is temporarily accelerated between the plates and then immediately slows down, this helps to aggregate very fine particles that can settle as the flow exits the plates.

Structures inclined between 45° and 60° may allow gravity drainage of accumulated solids, but shallower angles of inclination typically require periodic draining and cleaning. Tube settlers may allow the use of a smaller clarifier and may enable finer particles to be separated with residence times less than 10 minutes. Typically such structures are used for difficult-to-treat waters, especially those containing colloidal materials.

Tube settlers capture the fine particles allowing the larger particles to travel to the bottom of the clarifier in a more uniform form. The fine particles then build up into a larger mass which then slides down the tube channels. The reduction in solids present in the outflow allows a reduction in the clarifier footprint when designing. Tubes made of PVC plastic are a minor cost in clarifier design improvements and may lead to an increase of operating rate of 2 to 4 times.

Operation

In order to maintain and promote the proper processing of a clarifier, it is important to remove any corrosive, reactive and polymerisable component first, or any material that may foul the outlet stream of water to avoid any unwanted side reactions, changes in the product or cause damage to any of the water treatment equipment. This is done by routine inspections and the frequent cleaning of the quiescent zones and the inlet and outlet areas of the clarifier in order to ascertain the extent of sediment build up and to clean and remove any scouring, litter, weeds or debris that may have accumulated over time.

Water being introduced into the clarifier should be controlled to reduce the velocity of the inlet flow. Reducing the velocity maximizes the hydraulic retention time inside the clarifier for sedimentation and helps to avoid excessive turbulence and mixing; thereby promoting the effective settling of the suspended particles. To further discourage the overt mixing within the clarifier and increase the retention time allowed for the particles to settle, the inlet flow should also be distributed evenly across the entire cross section of the settling zone inside the clarifier, where the volume is maintained at 37.7 percent capacity.

The sludge formed from the settled particles at the bottom of each clarifier, if left for an extended period of time, may become gluey and viscous, causing difficulties in its removal. This formation of sludge, promotes anaerobic conditions and a healthy environment for the growth of bacteria. This can cause the resuspension of particles by gases and the release of dissolved nutrients throughout the water fluid, reducing the effectiveness of the clarifier. Major health issues and problems can also occur further down the track of the water purification system or hinder the health of the fish found downstream of the clarifier.

Settling Basin

One of the methods implemented to address sediment loading, a settling basin or settling pond, have become widespread across the valley as a quick and simple way to remove sediments out of return flows. Settling basins simply slow down return flows, allowing sediments to settle out of the water before the water returns to the irrigation delivery system or other water body.

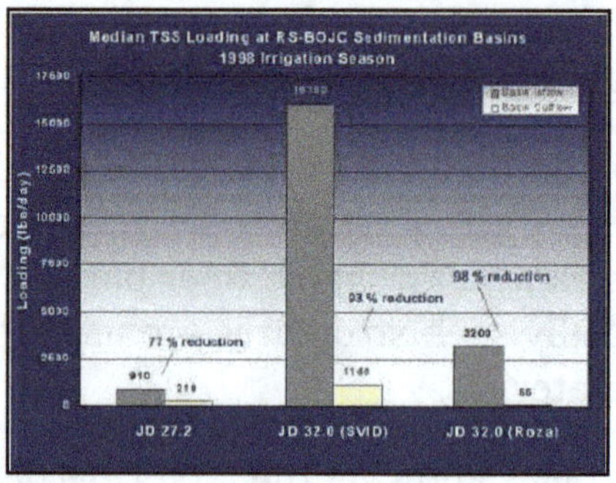

This graph shows the decrease in sediment loading over time in settling basins. Each basin's performance will vary depending on detention time, volume, and size. These variables will provide for varying results, but the bottom line, as evidence in the graph above, settling basins work very well removing sedimentation from return flows.

Pictured below is an example of an on-farm settling basin. In this application, return flow from the farm is collected in a settling basin (foreground) before returning to the irrigation canal (background).

Range of Applications

Settling basins are used as a separation mechanism to eliminate rejected products (i.e. waste solids management strategies) of a specified size and quantity in various fields, such as aquaculture, mining, dairy, food processing, alcohol production and wine making. Regular draining and desilting of settling basins is required to maintain satisfactory performance.

Aquaculture

All materials not removed from the system during harvesting are categorized as wastes including uneaten feed, excreta, chemicals and therapeutics, dead and moribund fish, escaped fish and pathogens. Settling basins in the field are simple ponds dug downstream of the farm to optimally remove suspended solids effectively, produce clarified effluent, and accumulate and thicken sludge to minimal volume. If impairment occurs in any of these functions, this might have a great impact on pond performance, which could lead to damaging the effectiveness of the process.

Mining

Wastewater produced by mining industries contribute to the acidity, suspended material and dissolved heavy metal ions in the aquatic environment, causing environmental problems for biological life and discoloration of the receiving waters. The application of settling basins by the Coeur d'Alene mining district of northern Idaho, United States, globally known to produce lead, zinc and silver, to treat wastewater has greatly improved the quality of water discharge from mining operations.

Dairy Waste

By reducing flow velocity to limit solids being transported along with fast flowing liquid, separation can occur. Approximately 35% – 60% of the solids is removed from dilute liquid slurry, with 10 minutes detention time, with a common detention time of 30 to 60 minutes. Due to the inadequate consideration of critical design criteria, most settling basins built were oversized and had low efficiency.

Settling basins used in dairy production reduce the nutrient-loading on a vegetative filter strip from lot runoff, thus decreasing the required lagoon volume for a new facility. Moreover, settling basins are useful to remove unwanted solid materials, such as hay, straw and feathers from the waste stream before flowing to the lagoon, aids to reduce smell and avoid crust formation on the lagoon surface. A baffle may be used to retain the floating solids removed. There are two types of settling basins, based on the method of removing solids. With one type, the solids are removed mechanically (after the free water has drained away), usually with a front-end or skid-steer loader. The other type uses hydraulic (pump) removal of the solids. Typically, pumping is initiated when the basin is half full of solids and the remainder is water. Vigorous agitation is needed to mix the liquid and the solids, preferably by propeller-type agitators or pumps with agitation nozzles.

There are two types of settling basins, based on the method of removing solids. With one type, the solids are removed mechanically (after the free water has drained away), usually with a front-end or skid-steer loader. The depth of accumulated solids should not exceed 1.5 feet. The other type uses hydraulic (pump) removal of the solids. Typically, pumping is initiated when the basin is half full of solids and the remainder is water. Vigorous agitation is needed to mix the liquid and the solids, preferably by propeller-type agitators or pumps with agitation nozzles.

Settling basins may be either concrete or earthen structures. For concrete basins, a common recommendation is a minimum depth of 2 feet plus the depth required for solids storage. Figure shows a typical concrete settling basin. Earthen structures may be compact basins, settling terraces, settling diversion terraces or settling channels. Figure shows a typical earthen settling basin. Earthen basins to be cleaned with loaders are usually designed to be shallow (not more than 3 feet deep) and to cover a large area. Earthen settling basins should have a concrete entrance ramp and a concrete runway on the bottom to allow entry of equipment for solids removal. Figure shows a duplex concrete settling basin that allows one side to receive effluent while the other side is thoroughly dewatered and the solids removed.

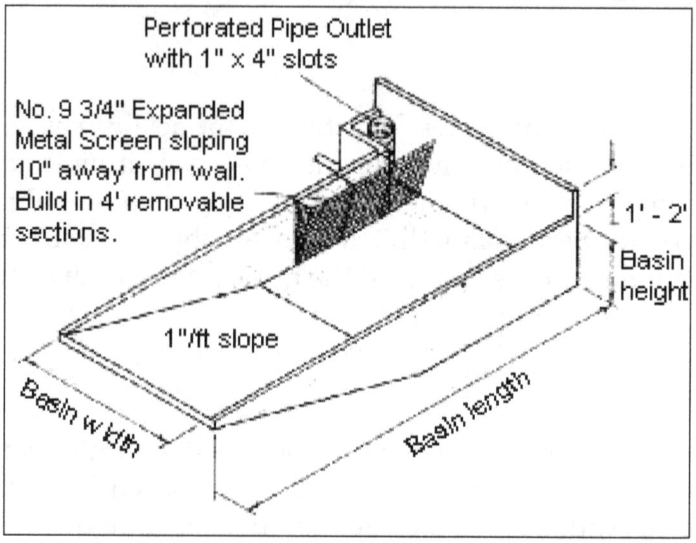

A typical concrete settling basin designed for mechanical removal of solids (from MWPS18).

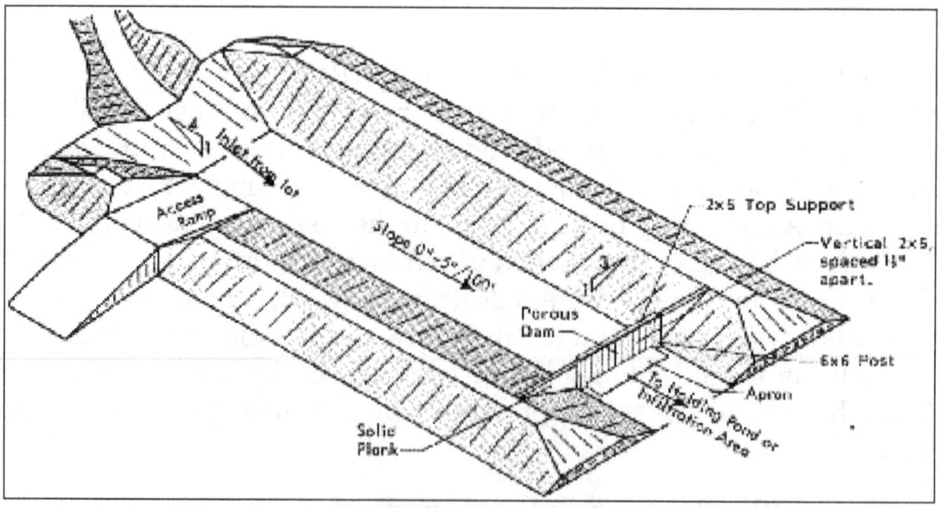

A typical earthen settling basin designed for mechanical removal of solids (from MWPS18).

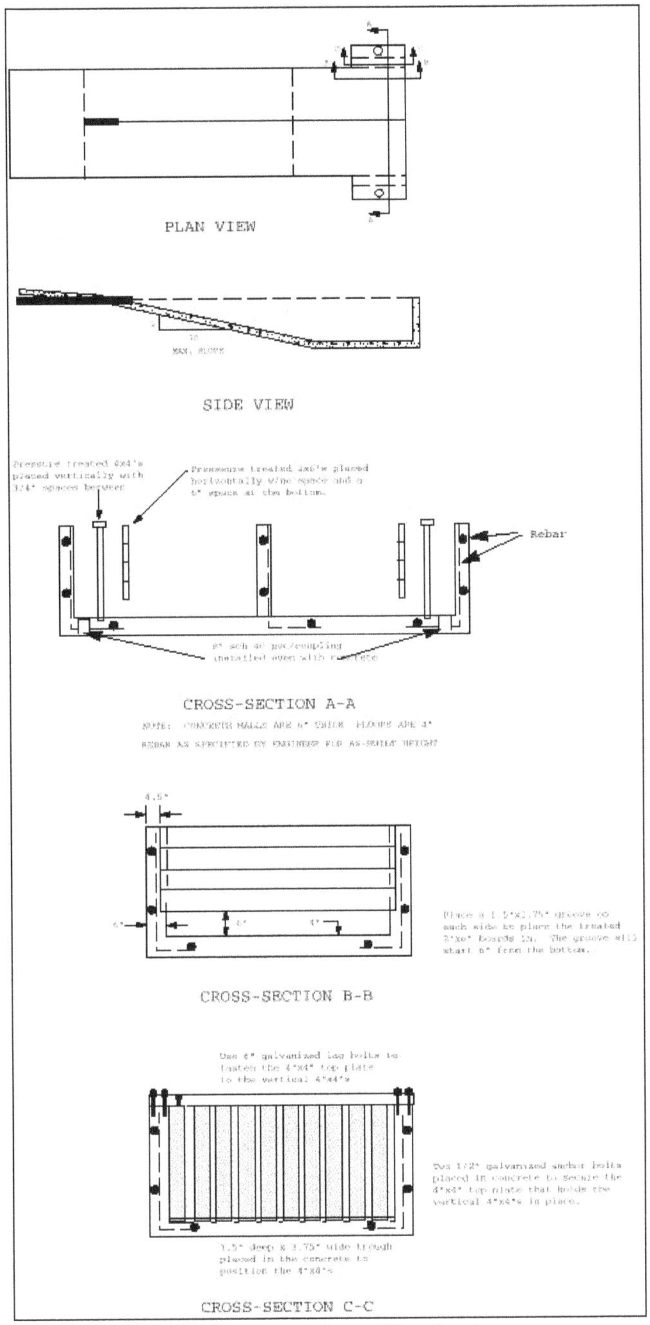

Duplex concrete settling basin.

A settling terrace, a settling diversion terrace or a settling channel is a wide, shallow, gently sloping, flat-bottomed waterway in which runoff solids settle due to low velocity. The channel is sometimes grassed to improve settling and reduce erosion. Grass may not survive in the channel and may make cleaning more difficult. Grass should be maintained on the sideslopes, if possible. Solids should be removed annually, or more often if required, to maintain capacity. Berm tops should be at least 2 feet wide to maintain the design height and at least 12 feet wide for vehicle traffic.

In Missouri's humid climate, inadequate drying of the solids and the channel may limit the usefulness of earthen settling terraces and channels.

Design of Concrete Settling Basins

An example will illustrate the design of a concrete settling basin. A blank form is included for your calculations.

Example

Design a concrete settling basin for a dirt lot, 100 feet by 200 feet, on a 6-percent slope. The solids will be pumped out at 6-month intervals. The basin length is to be four times the basin width. The location is in the northeast corner of Missouri. Design using peak runoff rate for a 1-year per 10-year storm from Table.

Table: Peak rates of runoff to be expected from watersheds in Missouri (Q).

Acres	One year per 10 (cubic feet per second per acre)	One year per 25[1] (cubic feet per second per acre)
0 to 1	5.5	0.2
1 to 2	4.6	6.0
2 to 3	4.2	5.5
3 to 4	3.0	5.2
4 to 20[2]	2.8	5.1

[1]One year per 10 is sufficient in most design situations. However, if a settling terrace is adjacent to a stream, use one year per 25.

[2]For areas greater than 20 acres, refer to MU publication G1518, Table 1.

Open Lot Area Draining into the Basin

Lot length 200 feet x lot width 100 feet = lot area 20,000 square feet

Inflow rate into settling basin

Q = peak runoff rate in cubic feet per second per acre
L = location factor
T = topographic factor
Q_T = inflow rate into settling basin, cubic feet per hour

QT = Q x L x T x square feet lot area x 0.0411 = cubic feet per hour

QT = 5.5 cubic feet per second per acre (Q) x 0.96 (L) x 0.92 (T) x 20,000 square feet x 0.0411 = 3,993 cubic feet per hour

If inflow arrives at settling basin via a sewer pipe, estimate QT from the size of the sewer pipe in Table.

QT = gpm (Table) x 8 = cubic feet per hour

If inflow arrives at settling basin by other means, explain (i.e., a dairy flush alley discharging into settling basin).

Estimate Inflow Rate

Q_T = estimated gpm x 8 = cubic feet per hour

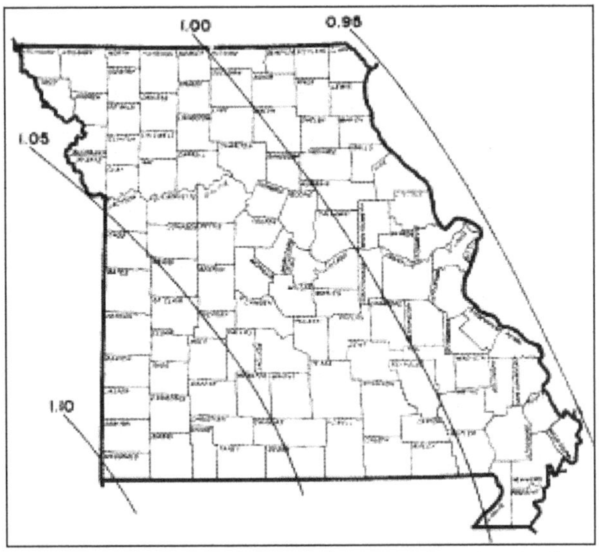

Location factors (L) from G1518, Estimating Peak Rates of Runoff From Small Watersheds.

Table: Topographic factors for different land slopes.[1]

Average land slope	Topographic factor (T)
1 percent	0.65
2 percent	0.72
3 percent	0.78
4 percent	0.83
5 percent	0.88
6 percent	0.92
7 percent	0.96
8 percent	1.00
9 percent	1.04
10 percent	1.07
12 percent	1.14
14 percent	1.20

[1]Obtained from MU publication G1518, *Estimating Peak Rates of Runoff from Small Watersheds.*

Table: Diameter, slope and expected flow rates for sewer lines.

Diameter	Slope	Expected flow rate
4 inches	0.02 foot per foot	115 GPM
6 inches	0.013 foot per foot	255 GPM
8 inches	0.009 foot per foot	460 GPM
10 inches	0.007 foot per foot	750 GPM

Surface Area (SA) of Settling Basin

SA (square feet)	=	Q_T (cubic feet per hour)/ 4 cubic feet per hour per square feet	=	3,993 cubic feet per hour/ 4 cubic feet per hour per square feet	=	998 square feet

Basin Dimensions

Design basin for length = 3 to 5 times basin width.

Basin width = $[SA/R]^{0.5}$

SA = basin surface area, square feet (from item 3)
R = length-width ratio = 4

width = $[998 \text{ square feet (SA)} \div 4 \text{ (R)}]^{0.5}$ = 15.8 feet

length = 16 feet (width) x 4 (R) = 64 feet

Basin Overflow

Provide a rectangular overflow weir at downstream end of basin. Weir height = 6 inches. Maximum weir length = width of settling basin. Minimum weir length, feet = $Q_T/1{,}250$. If a riser pipe is used as an overflow device instead of a rectangular weir, riser pipe diameter, inches = $Q_T/274$. Do not use a riser pipe smaller than 6 inches in diameter.

Minimum rectangular weir length = $Q_T/1{,}250$ = 3,993 /1,250 = 3.2 feet

Minimum riser pipe diameter = $Q_T/274$ = 3,993 /274 = 14.6 inches

Basin Depth

Use the following as a guide for volume of solids and calculate depth required for desired storage period.

Dirt lots = 2,800 cubic feet per acre-yr

Concrete lots and confinement buildings = 0.5 times manure production volume.

Table: Approximate daily manure production.

Animal	Weight	Cubic feet per day (liquid + solid)
Dairy	1,000	1.32
Beef (a) Concentration ration (b) Corn silage (c) 25 percent haylage (d) 50 percent haylage	1,000 1,000 1,000 1,000	0.35 0.45 0.34 0.43
Swine, finish	1,000	1.10
Sow + litter	1,000	1.44
Gestation sow	1,000	0.55

Boar	1,000	0.55
Sheep	1,000	0.62
Poultry (a) Layers (b) Broilers	1,000 1,000 1,000	0.88 1.20
Horse	1,000	1.50

If solids are to be removed from basin by pumping, design basin to hold an equal volume of water above settled solids. Solids must be diluted and agitated for pumping. If solids are to be removed mechanically (i.e., front-end loader), provide concrete entrance to settling basin with at least 10:1 slope (20:1 slope preferred). Additional dewatering via a hardware cloth dam and/or a perforated riser pipe is desirable for mechanical removal of settled solids.

Indicate desired storage period, days = 182

(Lot acres = 100 feet x 200 feet/43,560 = 0.46 acres)

$$\text{Basin depth (dirt lot)} = \frac{2{,}800 \text{ x } 0.46 \text{ acres lot x } 182 \text{ days storage x } 2^*}{365 \text{ days per yr x } 998 \text{ square feet surface area (item 3)}} = 1.3 \text{ feet}$$

*Multiply by 1 for mechanical removal or by 2 for pumping.

Basin depth, BD_C (concrete lot or confinement building)

$$BD_C = \frac{0.5 \text{ x \_\_ cubic feet per day manure prod (Table) x \_\_ days storage x } (1^* \text{ or } 2^*)}{\text{\_\_\_\_\_\_ square feet surface area (item 3)}} = \text{\_\_\_\_\_\_ feet}$$

*Multiply by 1 for mechanical removal or by 2 for pumping.

Note

When settling basin discharges into a lagoon, the size of the lagoon may be reduced as follows:
Design volume (with settling basin) = Design volume (without settling basin) x 0.5
Manure storage volume (with settling basin) = Manure storage volume (without settling basin) x 0.5

Minimum design storage period is 90 days when the lagoon design volume is reduced by 50 percent as noted above. Less storage may be used if the lagoon design volume is based on 100 percent loading.

Basin Outlets

Various types of basin outlets are used to drain liquids from the full depth of basins and allow the solids to dewater. The porous plank dam ahead of either a perforated or a slotted riser pipe are frequently used outlets. Manure tends to plug even large openings in outlets. Unplugging is required frequently. A hoe may be used to scrape solids off of openings. Also, a slanted expanded metal or quarry screen with 1-inch to 1.5-inch openings may be used around the outlet to increase the screening area and reduce clogging.

Porous Dams

Porous dams may be made of welded wire fabric, expanded metal mesh or spaced boards. Porous dams may be used to dewater settling basins or to remove large solids that tend to cause excessive clogging of the openings in perforated pipe outlets. Dams constructed with spaced boards usually have 0.75-inch spaces between the boards. The boards usually range from 2-by-6s to 2-by-12s. Expanded metal and welded wire fabric have even greater open areas. Due to the large open area in a porous dam, little design is required. As a general rule, the open area in a porous dam should be twice the area of the perforations in the riser pipe it precedes. As a practical matter, a porous dam 4 feet long or more, should suffice for the common sizes of outlet pipes. In some applications, there is no outlet pipe and the porous dam forms one wall of the settling/storage basin.

Perforated Pipe Outlets

Material for perforated pipe is usually PVC plastic, galvanized steel or concrete. Perforations can be 5/8-inch to 1-inch diameter holes or 1-inch by 4-inch slots. The outlet is sized to match the anticipated flow rates in order to assure adequate detention time. Flow rate is controlled by the amount of open area (slots or holes) in the pipe. Table gives opening requirements for perforated pipes.

Table: Riser pipe outlet design for settling basins (from MWPS18).

Based on $Q = (C)(A)(2gh)^{0.5}$, where Q = flow rate in cfs; C = slot constant, assumed to be 0.61; A = open slot area in square feet; g = 32.174 feet per second^2; h = head on openings in feet. The pipe height is divided into 0.5-foot increments. The head on all the slots in the first 0.5-foot increment is assumed to be 0.25-foot. The head on subsequent 0.5-foot increments increase at 0.5-foot increments.

Open slot area per foot of pipe height, in ^{2} per foot	**Head, feet**							
	0.5	**1.0**	**1.5**	**2.0**	**2.5**	**3.0**	**3.5**	**4.0**
	cfs							
4	0.034	0.093	0.169	0.259	0.361	0.473	0.596	0.728
6	0.051	0.139	0.253	0.388	0.541	0.710	0.894	1.091
8	0.068	0.186	0.338	0.518	0.721	0.947	1.192	1.455
10	0.085	0.232	0.422	0.647	0.902	1.183	1.480	1.819
12	0.102	0.279	0.507	0.776	1.082	1.420	1.788	2.183
14	0.119	0.325	0.591	0.906	1.262	1.657	2.086	2.546
16	0.136	0.371	0.675	1.035	1.443	1.894	2.384	2.910
18	0.153	0.418	0.760	1.164	1.623	2.130	2.682	3.274
20	0.170	0.464	0.844	1.294	1.803	2.367	3.890	3.638
22	0.187	0.511	0.929	1.423	1.984	2.604	3.277	4.001
24	0.204	0.557	1.013	1.542	2.164	2.840	3.575	4.365
26	0.221	0.603	1.097	1.682	2.344	3.077	3.873	4.729
28	0.238	0.650	1.182	1.811	2.525	3.314	4.171	5.093
30	0.255	0.696	1.266	1.940	2.705	3.550	4.469	5.456
32	0.272	0.743	1.351	2.070	2.885	3.787	4.767	5.820
34	0.289	0.789	1.435	2.199	3.066	4.024	5.065	6.184

36	0.306	0.836	1.519	2.329	3.246	4.260	5.363	6.548
38	0.323	0.882	1.604	2.458	3.426	4.497	5.661	6.911
40	0.340	0.928	1.688	2.587	3.607	4.734	5.959	7.275

Example

Design a basin outlet to allow outflow to equal peak flow rate off the lot in Example 1 when the basin is full. The inflow rate in Example 1 is 3,993 cubic feet per hour and the depth is 1.3 feet.

- For a perforated pipe riser, determine the required open area per foot of pipe height from Table 5. Outflow = 3,993 cubic feet per hour = 1.1 cubic feet per second. By interpolation in Table 5, we find that with 1.3 feet of head the required opening area in the riser pipe is 32 in^2 per foot of pipe height.
- Size the outlet pipe from the data in Table 3. Outflow in gpm = 1.1 cubic feet per second x 450 gpm per cfs = 495 gpm. From Table 3, an 8-inch pipe at 0.009 slope will carry 460 gpm and a 10-inch pipe at 0.007 slope will carry 750 gpm. Depending on slope, use an 8-inch or 10-inch pipe (an 8-inch pipe will carry 1.1 cfs at slopes greater than 1 percent, Ref. Figure 4.5b in MWPS18).

Check on the time to withdraw the 9,167 cubic feet of runoff from the 25-year, 24-hour storm at 1.1 cfs.

Time in hours = 9,167 cubic feet ÷ (1.1 cubic feet per second x 3,600 second per hour) = 2.3 hours at maximum head.

Design of Earthen Settling Basins

To meet approval by the Missouri Department of Natural Resources, earthen basins must be built as follows. Berms shall have minimum slopes of 3:1. If solids are to be removed using mechanical equipment, a concrete pad shall be installed in the bottom of the basin and a concrete access ramp with a maximum slope of 10 percent shall be provided. If the settled solids are to be removed by pumping, the basin must be designed to contain an equal volume of water above the solids to allow for agitation and dilution of the solids. Access points for the mixing equipment must be indicated on the construction drawings.

Design of Settling Channels

A settling diversion terrace, settling terrace or a settling channel is a wide, shallow, gently sloping, flat-bottomed channel, in which suspended solids contained in runoff water are settled out. A settling channel may be either of earthen or concrete construction. The settling channel may be grassed to improve settling and reduce erosion. Runoff water from the channel is stored in a lagoon or storage pond. Wastes settled from the runoff are allowed to dry before removal with mechanical equipment, typically with a tractor and front-end loader. Usually, solids are removed from the channel once per year or when accumulated solids reduce the settling ability of the channel.

Sideslopes for settling channels usually range from 3:1 to 4:1, depending on soil properties. The bottom slope of the channel should be between 0.1 percent and 0.3 percent to maintain low velocities and rapid settling.

Sewage Treatment

Section of a wastewater treatment plant.

Sewage treatment, or domestic wastewater treatment, is the process of removing contaminants from wastewater and household sewage, both runoff (effluents) and domestic. It includes physical, chemical and biological processes to remove physical, chemical and biological contaminants. Its objective is to produce a waste stream (or treated effluent) and a solid waste or sludge suitable for discharge or reuse back into the environment. This material is often inadvertently contaminated with many toxic organic and inorganic compounds.

Sewage is created by residences, institutions, hospitals and commercial and industrial establishments. It can be treated close to where it is created (in septic tanks, biofilters or aerobic treatment systems), or collected and transported via a network of pipes and pump stations to a municipal treatment plant. Sewage collection and treatment is typically subject to local, state and federal regulations and standards. Industrial sources of wastewater often require specialized treatment processes.

The sewage treatment involves three stages, called *primary*, *secondary*, and *tertiary treatment*. First, the solids are separated from the wastewater stream. Then, dissolved biological matter is progressively converted into a solid mass by using indigenous, water-borne micro-organisms. Finally, the biological solids are neutralized, then disposed of or re-used, and the treated water may be disinfected chemically or physically (for example by lagoons and micro-filtration). The final effluent can be discharged into a stream, river, bay, lagoon or wetland, or it can be used for the irrigation of a golf course, green way or park. If it is sufficiently clean, it can also be used for groundwater recharge or agricultural purposes.

Description

Raw influent (sewage) includes household waste liquid from toilets, baths, showers, kitchens, sinks, and so forth that is disposed of via sewers. In many areas, sewage also includes liquid waste from industry and commerce.

The separation and draining of household waste into greywater and blackwater is becoming more common in the developed world, with greywater being permitted to be used for watering plants or recycled for flushing toilets. A lot of sewage also includes some surface water from roofs or hard-standing areas. Municipal wastewater therefore includes residential, commercial, and industrial liquid waste discharges, and may include stormwater runoff. Sewage systems capable of handling stormwater are known as combined systems or combined sewers. Such systems are usually avoided since they complicate and thereby reduce the efficiency of sewage treatment plants owing to their seasonality. The variability in flow also leads to often larger than necessary, and subsequently more expensive, treatment facilities. In addition, heavy storms that contribute more flows than the treatment plant can handle may overwhelm the sewage treatment system, causing a spill or overflow (called a combined sewer overflow, or CSO, in the United States). It is preferable to have a separate storm drain system for stormwater in areas that are developed with sewer systems.

As rainfall runs over the surface of roofs and the ground, it may pick up various contaminants including soil particles and other sediment, heavy metals, organic compounds, animal waste, and oil and grease. Some jurisdictions require stormwater to receive some level of treatment before being discharged directly into waterways. Examples of treatment processes used for stormwater include sedimentation basins, wetlands, buried concrete vaults with various kinds of filters, and vortex separators (to remove coarse solids).

The site where the raw wastewater is processed before it is discharged back to the environment is called a wastewater treatment plant (WWTP). The order and types of mechanical, chemical and biological systems that comprise the wastewater treatment plant are typically the same for most developed countries:

- Mechanical treatment:
    - Influx (Influent).
    - Removal of large objects.
    - Removal of sand and grit.
    - Pre-precipitation.
- Biological treatment:
    - Oxidation bed (oxidizing bed) or aeration system.
    - Post precipitation.
- Chemical treatment this step is usually combined with settling and other processes to remove solids, such as filtration. The combination is referred to in the United States as physical chemical treatment.

Primary treatment removes the materials that can be easily collected from the raw wastewater and disposed of. The typical materials that are removed during primary treatment include fats, oils, and greases (also referred to as FOG), sand, gravels and rocks (also referred to as grit), larger settleable solids and floating materials (such as rags and flushed feminine hygiene products). This step is done entirely with machinery.

Process Flow Diagram for a Typical Large-Scale Treatment Plant

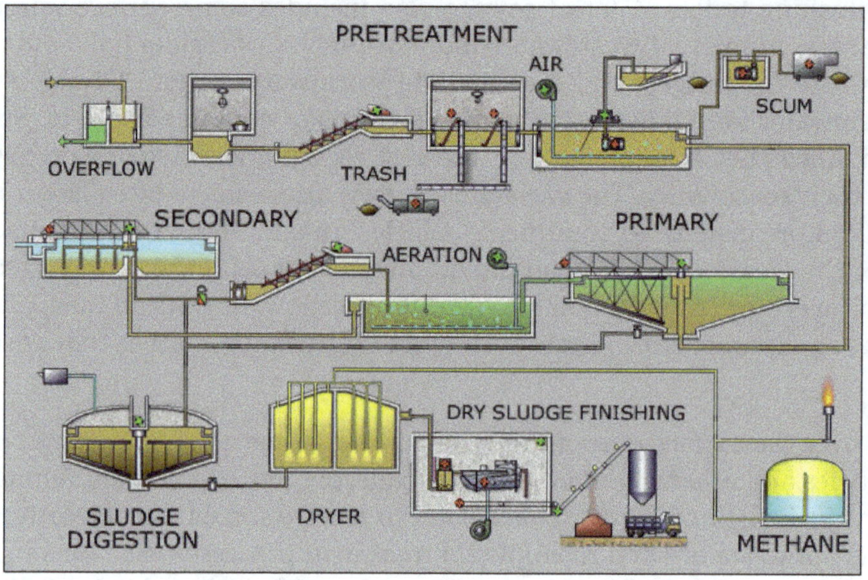

Process Flow Diagram for a Typical Treatment Plant via Subsurface Flow Constructed Wetlands (SFCW).

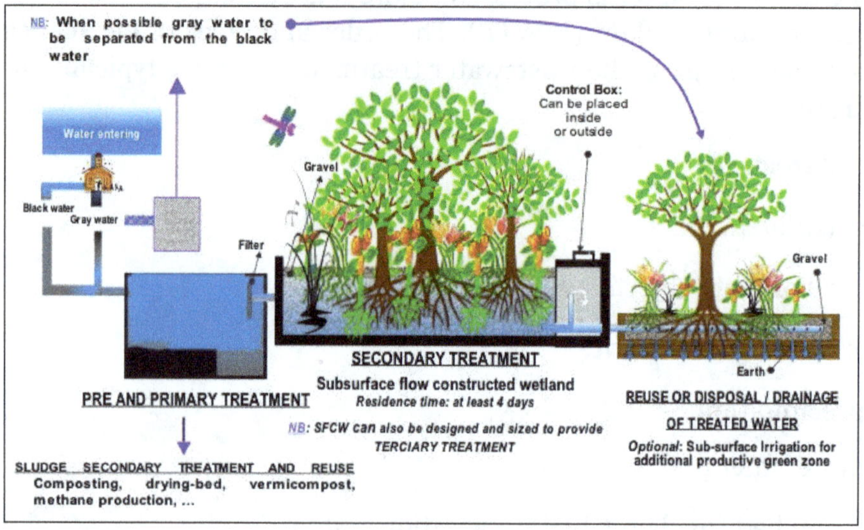

Primary treatment

Removal of Large Objects from Influent Sewage

In primary treatment, the influent sewage water is strained to remove all large objects that are deposited in the sewer system, such as rags, sticks, tampons, cans, fruit, etc. This is most commonly done with a manual or automated mechanically raked bar screen. The raking action of a mechanical bar screen is typically paced according to the accumulation on the bar screens and/or flow rate. The bar screen is used because large solids can damage or clog the equipment used later in the sewage treatment plant. The solids are collected in a dumpster and later disposed in a landfill.

Primary treatment also typically includes a sand or grit channel or chamber where the velocity of the incoming wastewater is carefully controlled to allow sand grit and stones to settle, while keeping the majority of the suspended organic material in the water column. This equipment is called a degritter or sand catcher. Sand, grit, and stones need to be removed early in the process to avoid damage to pumps and other equipment in the remaining treatment stages. Sometimes there is a sand washer (grit classifier) followed by a conveyor that transports the sand to a container for disposal. The contents from the sand catcher may be fed into the incinerator in a sludge processing plant, but in many cases, the sand and grit is sent to a landfill.

Empty sedimentation tank at the treatment plant in Merchtem.

Sedimentation

Many plants have a sedimentation stage where the sewage is allowed to pass slowly through large tanks, commonly called "primary clarifiers" or "primary sedimentation tanks." The tanks are large enough that sludge can settle and floating material such as grease and oils can rise to the surface and be skimmed off. The main purpose of the primary clarification stage is to produce both a generally homogeneous liquid capable of being treated biologically and a sludge that can be separately treated or processed. Primary settling tanks are usually equipped with mechanically driven scrapers that continually drive the collected sludge towards a hopper in the base of the tank from where it can be pumped to further sludge treatment stages.

Secondary Treatment

Secondary treatment is designed to substantially degrade the biological content of the sewage such as are derived from human waste, food waste, soaps and detergent. The majority of municipal plants treat the settled sewage liquor using aerobic biological processes. For this to be effective, the biota require both oxygen and a substrate on which to live. There are a number of ways in which this is done. In all these methods, the bacteria and protozoa consume biodegradable soluble organic contaminants (e.g. sugars, fats, organic short-chain carbon molecules, etc.) and bind much of the less soluble fractions into floc. Secondary treatment systems are classified as fixed film or suspended growth. Fixed-film treatment process including trickling filter and rotating biological contactors where the biomass grows on media and the sewage passes over its surface. In suspended growth systems—such as activated sludge—the biomass is well mixed with the sewage and can be operated in a smaller space than fixed-film systems that treat the same amount of water.

However, fixed-film systems are more able to cope with drastic changes in the amount of biological material and can provide higher removal rates for organic material and suspended solids than suspended growth systems.

Roughing filters are intended to treat particularly strong or variable organic loads, typically industrial, to allow them to then be treated by conventional secondary treatment processes. Characteristics include typically tall, circular filters filled with open synthetic filter media to which wastewater is applied at a relatively high rate. They are designed to allow high hydraulic loading and a high flow-through of air. On larger installations, air is forced through the media using blowers. The resultant wastewater is usually within the normal range for conventional treatment processes.

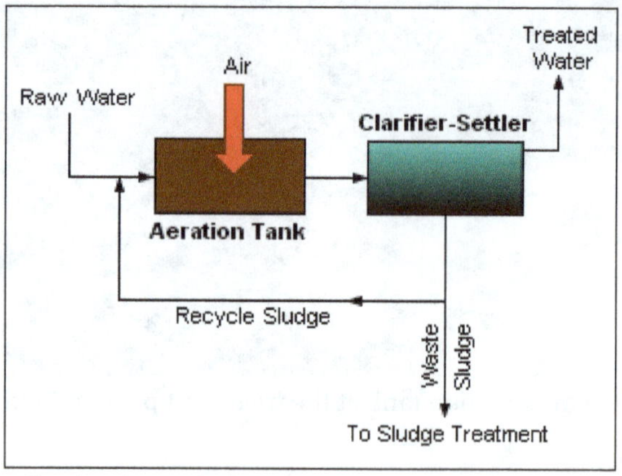

A generalized, schematic diagram of an activated sludge process.

Activated Sludge

In general, activated sludge plants encompass a variety of mechanisms and processes that use dissolved oxygen to promote the growth of biological floc that substantially removes organic material.

The process traps particulate material and can, under ideal conditions, convert ammonia to nitrite and nitrate and ultimately to nitrogen gas.

Surface-aerated Basins

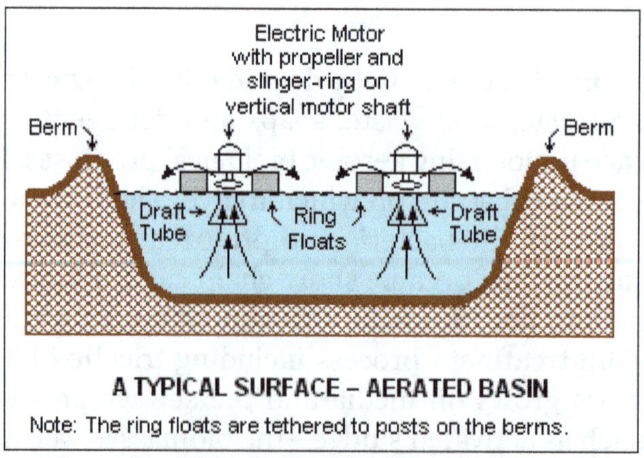

A Typical Surface-Aerated Basin (using motor-driven floating aerators).

Most biological oxidation processes for treating industrial wastewaters have in common the use of oxygen (or air) and microbial action. Surface-aerated basins achieve 80 to 90 percent removal of Biochemical Oxygen Demand with retention times of 1 to 10 days. The basins may range in depth from 1.5 to 5.0 meters and use motor-driven aerators floating on the surface of the wastewater.

In an aerated basin system, the aerators provide two functions: they transfer air into the basins required by the biological oxidation reactions, and they provide the mixing required for dispersing the air and for contacting the reactants (that is, oxygen, wastewater and microbes). Typically, the floating surface aerators are rated to deliver the amount of air equivalent to 1.8 to 2.7 kg O_2/kW•h. However, they do not provide as good mixing as is normally achieved in activated sludge systems and therefore aerated basins do not achieve the same performance level as activated sludge units.

Biological oxidation processes are sensitive to temperature and, between 0 °C and 40 °C, the rate of biological reactions increase with temperature. Most surface aerated vessels operate at between 4 °C and 32 °C.

Fluidized Bed Reactors

The carbon absorption following biological treatment is particularly effective in reducing both the BOD and COD to low levels. A fluidized bed reactor is a combination of the most common stirred tank packed bed, continuous flow reactors. It is very important to chemical engineering because of its excellent heat and mass transfer characteristics. In a fluidized bed reactor, the substrate is passed upward through the immobilized enzyme bed at a high velocity to lift the particles. However the velocity must not be so high that the enzymes are swept away from the reactor entirely. This causes low mixing; these type of reactors are highly suitable for the exothermic reactions. It is most often applied in immobilized enzyme catalysis.

Filter Beds (Oxidizing Beds)

In older plants and plants receiving more variable loads, trickling filter beds are used where the settled sewage liquor is spread onto the surface of a deep bed made up of coke (carbonized coal), limestone chips or specially fabricated plastic media. Such media must have high surface areas to support the biofilms that form. The liquor is distributed through perforated rotating arms radiating from a central pivot. The distributed liquor trickles through this bed and is collected in drains at the base. These drains also provide a source of air which percolates up through the bed, keeping it aerobic. Biological films of bacteria, protozoa and fungi form on the media's surfaces and eat or otherwise reduce the organic content. This biofilm is grazed by insect larvae and worms which help maintain an optimal thickness. Overloading of beds increases the thickness of the film leading to clogging of the filter media and ponding on the surface.

Biological Aerated Filters

Biological Aerated (or Anoxic) Filter (BAF) or Biofilters combine filtration with biological carbon reduction, nitrification or denitrification. BAF usually includes a reactor filled with a filter media. The media is either in suspension or supported by a gravel layer at the foot of the filter. The dual purpose of this media is to support highly active biomass that is attached to it and to filter suspended solids. Carbon reduction and ammonia conversion occurs in aerobic mode and sometime achieved in a single reactor while nitrate conversion occurs in anoxic mode. BAF is operated either in upflow or downflow configuration depending on design specified by manufacturer.

Secondary Sedimentation tank at a rural treatment plant.

Membrane Bioreactors

Membrane bioreactors (MBR) combines activated sludge treatment with a membrane liquid-solid separation process. The membrane component uses low pressure microfiltration or ultra filtration membranes and eliminates the need for clarification and tertiary filtration. The membranes are typically immersed in the aeration tank (however, some applications utilize a separate membrane tank). One of the key benefits of a membrane bioreactor system is that it effectively overcomes the limitations associated with poor settling of sludge in conventional activated sludge (CAS) processes. The technology permits bioreactor operation with considerably higher mixed liquor suspended solids (MLSS) concentration than CAS systems, which are limited by sludge settling. The process is typically operated at MLSS in the range of 8,000–12,000 mg/L, while CAS are operated in the range of 2,000–3,000 mg/L. The elevated biomass concentration in the membrane bioreactor process allows for very effective removal of both soluble and particulate biodegradable materials at higher loading rates. Thus increased Sludge Retention Times (SRTs)—usually exceeding 15 days—ensure complete nitrification even in extremely cold weather.

The cost of building and operating a MBR is usually higher than conventional wastewater treatment, however, as the technology has become increasingly popular and has gained wider acceptance throughout the industry, the life-cycle costs have been steadily decreasing. The small footprint of MBR systems, and the high quality effluent produced, makes them particularly useful for water reuse applications.

Secondary Sedimentation

The final step in the secondary treatment stage is to settle out the biological floc or filter material and produce sewage water containing very low levels of organic material and suspended matter.

Rotating Biological Contactors

Rotating biological contactors (RBCs) are mechanical secondary treatment systems, which are robust and capable of withstanding surges in organic load. RBCs were first installed in Germany in 1960 and have since been developed and refined into a reliable operating unit. The rotating disks support the growth of bacteria and micro-organisms present in the sewage, which breakdown and stabilize organic pollutants. To be successful, micro-organisms need both oxygen to live and food to grow. Oxygen is obtained from the atmosphere as the disks rotate. As the micro-organisms grow, they build up on the media until they are sloughed off due to shear forces provided by the rotating discs in the sewage. Effluent from the RBC is then passed through final clarifiers where the micro-organisms in suspension settle as a sludge. The sludge is withdrawn from the clarifier for further treatment.

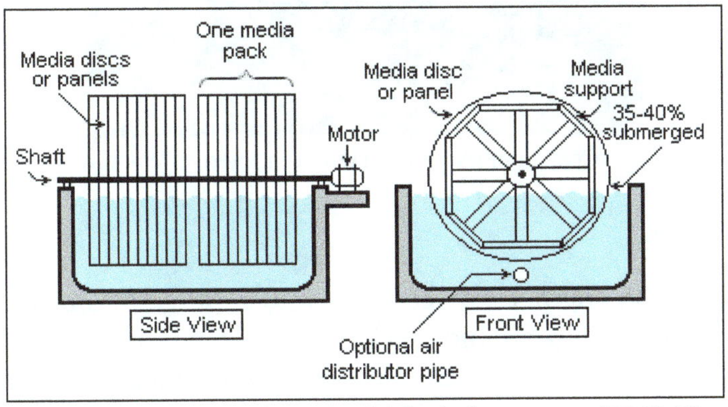

Schematic diagram of a typical rotating biological contactor (RBC). The treated effluent clarifier/settler is not included in the diagram.

A functionally similar biological filtering system has become popular as part of home aquarium filtration and purification. The aquarium water is drawn up out of the tank and then cascaded over a freely spinning corrugated fiber-mesh wheel before passing through a media filter and back into the aquarium. The spinning mesh wheel develops a biofilm coating of microorganisms that feed on the suspended wastes in the aquarium water and are also exposed to the atmosphere as the wheel rotates. This is especially good at removing waste urea and ammonia urinated into the aquarium water by the fish and other animals.

Tertiary Treatment

The purpose of tertiary treatment is to provide a final treatment stage to raise the effluent quality before it is discharged to the receiving environment (sea, river, lake, ground, etc.). More than one tertiary treatment process may be used at any treatment plant. If disinfection is practiced, it is always the final process. It is also called "effluent polishing."

Filtration

Sand filtration removes much of the residual suspended matter. Filtration over activated carbon removes residual toxins.

Lagooning

Lagooning provides settlement and further biological improvement through storage in large man-made ponds or lagoons. These lagoons are highly aerobic and colonization by native macrophytes, especially reeds, is often encouraged. Small filter feeding invertebrates such as Daphnia and species of Rotifera greatly assist in treatment by removing fine particulates.

A sewage treatment plant and lagoon in Everett, Washington.

Constructed Wetlands

Constructed wetlands include engineered reedbeds and a range of similar methodologies, all of which provide a high degree of aerobic biological improvement and can often be used instead of secondary treatment for small communities, also see phytoremediation. One example is a small reedbed used to clean the drainage from the elephants' enclosure at Chester Zoo in England.

Nutrient Removal

Wastewater may contain high levels of the nutrients nitrogen and phosphorus. Excessive release to the environment can lead to a build up of nutrients, called eutrophication, which can in turn encourage the overgrowth of weeds, algae, and cyanobacteria (blue-green algae). This may cause an algal bloom, a rapid growth in the population of algae. The algae numbers are unsustainable and eventually most of them die. The decomposition of the algae by bacteria uses up so much of oxygen in the water that most or all of the animals die, which creates more organic matter for the bacteria to decompose. In addition to causing deoxygenation, some algal species produce toxins that contaminate drinking water supplies. Different treatment processes are required to remove nitrogen and phosphorus.

Nitrogen Removal

The removal of nitrogen is effected through the biological oxidation of nitrogen from ammonia (nitrification) to nitrate, followed by denitrification, the reduction of nitrate to nitrogen gas. Nitrogen gas is released to the atmosphere and thus removed from the water.

Nitrification itself is a two-step aerobic process, each step facilitated by a different type of bacteria. The oxidation of ammonia (NH_3) to nitrite (NO_2^-) is most often facilitated by *Nitrosomonas* spp. (nitroso referring to the formation of a nitroso functional group). Nitrite oxidation to nitrate (NO_3^-), though traditionally believed to be facilitated by *Nitrobacter* spp. (nitro referring the formation of a nitro functional group), is now known to be facilitated in the environment almost exclusively by *Nitrospira* spp.

Denitrification requires anoxic conditions to encourage the appropriate biological communities to form. It is facilitated by a wide diversity of bacteria. Sand filters, lagooning and reed beds can all be used to reduce nitrogen, but the activated sludge process (if designed well) can do the job the most easily. Since denitrification is the reduction of nitrate to dinitrogen gas, an electron donor is needed. This can be, depending on the wastewater, organic matter (from faeces), sulfide, or an added donor like methanol.

Sometimes the conversion of toxic ammonia to nitrate alone is referred to as tertiary treatment.

Phosphorus Removal

Phosphorus removal is important as it is a limiting nutrient for algae growth in many fresh water systems (for negative effects of algae see Nutrient removal). It is also particularly important for water reuse systems where high phosphorus concentrations may lead to fouling of downstream equipment such as reverse osmosis.

Phosphorus can be removed biologically in a process called enhanced biological phosphorus removal. In this process, specific bacteria, called polyphosphate accumulating organisms (PAOs), are selectively enriched and accumulate large quantities of phosphorus within their cells (up to 20 percent of their mass). When the biomass enriched in these bacteria is separated from the treated water, these biosolids have a high fertilizer value.

Phosphorus removal can also be achieved by chemical precipitation, usually with salts of iron (e.g. ferric chloride), aluminum (e.g. alum), or lime. This may lead to excessive sludge productions as hydroxides precipitates and the added chemicals can be expensive. Despite this, chemical phosphorus removal requires significantly smaller equipment footprint than biological removal, is easier to operate and is often more reliable than biological phosphorus removal.

Once removed, phosphorus, in the form of a phosphate rich sludge, may be land filled or, if in suitable condition, resold for use in fertilizer.

Disinfection

The purpose of disinfection in the treatment of wastewater is to substantially reduce the number of microorganisms in the water to be discharged back into the environment. The effectiveness of disinfection depends on the quality of the water being treated (e.g., cloudiness, pH, etc.), the type of disinfection being used, the disinfectant dosage (concentration and time), and other environmental variables. Cloudy water will be treated less successfully since solid matter can shield organisms, especially from ultraviolet light or if contact times are low. Generally, short contact times, low doses and high flows all militate against effective disinfection. Common methods of disinfection include ozone, chlorine, or ultraviolet light. Chloramine, which is used for drinking water, is not used in wastewater treatment because of its persistence.

Chlorination remains the most common form of wastewater disinfection in North America due to its low cost and long-term history of effectiveness. One disadvantage is that chlorination of residual organic material can generate chlorinated-organic compounds that may be carcinogenic or harmful to the environment. Residual chlorine or chloramines may also be capable of chlorinating organic material in the natural aquatic environment. Further, because residual chlorine is toxic to aquatic species, the treated effluent must also be chemically dechlorinated, adding to the complexity and cost of treatment.

Ultraviolet (UV) light can be used instead of chlorine, iodine, or other chemicals. Because no chemicals are used, the treated water has no adverse effect on organisms that later consume it, as may be the case with other methods. UV radiation causes damage to the genetic structure of bacteria, viruses, and other pathogens, making them incapable of reproduction. The key disadvantages of UV disinfection are the need for frequent lamp maintenance and replacement and the need for a highly treated effluent to ensure that the target microorganisms are not shielded from the UV radiation (i.e., any solids present in the treated effluent may protect microorganisms from the UV light). In the United Kingdom, light is becoming the most common means of disinfection because of the concerns about the impacts of chlorine in chlorinating residual organics in the wastewater and in chlorinating organics in the receiving water. Edmonton, Alberta, Canada also uses UV light for its water treatment.

Ozone O_3 is generated by passing oxygen O_2 through a high voltage potential resulting in a third oxygen atom becoming attached and forming O_3. Ozone is very unstable and reactive and oxidizes most organic material it comes in contact with, thereby destroying many pathogenic microorganisms. Ozone is considered to be safer than chlorine because, unlike chlorine which has to be stored on site (highly poisonous in the event of an accidental release), ozone is generated onsite as needed. Ozonation also produces fewer disinfection by-products than chlorination. A disadvantage of ozone disinfection is the high cost of the ozone generation equipment and the requirements for special operators.

Package Plants and Batch Reactors

In order to use less space, treat difficult waste, deal with intermittent flow or achieve higher environmental standards, a number of designs of hybrid treatment plants have been produced. Such plants often combine all or at least two stages of the three main treatment stages into one combined stage. In the UK, where a large number of sewage treatment plants serve small populations, package plants are a viable alternative to building discrete structures for each process stage.

One type of system that combines secondary treatment and settlement is the sequencing batch reactor (SBR). Typically, activated sludge is mixed with raw incoming sewage and mixed and aerated. The resultant mixture is then allowed to settle producing a high quality effluent. The settled sludge is run off and re-aerated before a proportion is returned to the head of the works. SBR plants are now being deployed in many parts of the world including North Liberty, Iowa, and Llanasa, North Wales.

The disadvantage of such processes is that precise control of timing, mixing and aeration is required. This precision is usually achieved by computer controls linked to many sensors in the plant. Such a complex, fragile system is unsuited to places where such controls may be unreliable, or poorly maintained, or where the power supply may be intermittent.

Package plants may be referred to as *high charged* or *low charged*. This refers to the way the biological load is processed. In high charged systems, the biological stage is presented with a high organic load and the combined floc and organic material is then oxygenated for a few hours before being charged again with a new load. In the low charged system the biological stage contains a low organic load and is combined with flocculate for a relatively long time.

Sludge Treatment and Disposal

The sludges accumulated in a wastewater treatment process must be treated and disposed of in a safe and effective manner. The purpose of digestion is to reduce the amount of organic matter and the number of disease-causing microorganisms present in the solids. The most common treatment options include anaerobic digestion, aerobic digestion, and composting.

Choice of a wastewater solid treatment method depends on the amount of solids generated and other site-specific conditions. However, in general, composting is most often applied to smaller-scale applications followed by aerobic digestion and then lastly anaerobic digestion for the larger-scale municipal applications.

Anaerobic Digestion

Anaerobic digestion is a bacterial process that is carried out in the absence of oxygen. The process can either be *thermophilic* digestion, in which sludge is fermented in tanks at a temperature of 55°C, or *mesophilic*, at a temperature of around 36°C. Though allowing shorter retention time (and thus smaller tanks), thermophilic digestion is more expensive in terms of energy consumption for heating the sludge.

One major feature of anaerobic digestion is the production of biogas, which can be used in generators for electricity production and/or in boilers for heating purposes.

Aerobic Digestion

Aerobic digestion is a bacterial process occurring in the presence of oxygen. Under aerobic conditions, bacteria rapidly consume organic matter and convert it into carbon dioxide. The operating costs used to be characteristically much greater for aerobic digestion because of the energy used by the blowers, pumps, and motors needed to add oxygen to the process. However, recent technological advances include non-electric aerated filter systems that use natural air currents for the aeration instead of electrically operated machinery. Aerobic digestion can also be achieved by using diffuser systems or jet aerators to oxidize the sludge.

Composting

Composting is also an aerobic process that involves mixing the sludge with sources of carbon such as sawdust, straw or wood chips. In the presence of oxygen, bacteria digest both the wastewater solids and the added carbon source and, in doing so, produce a large amount of heat.

Sludge Disposal

When a liquid sludge is produced, further treatment may be required to make it suitable for final disposal. Typically, sludges are thickened (dewatered) to reduce the volumes transported off-site for disposal. There is no process which completely eliminates the need to dispose of biosolids. There is, however, an additional step some cities are taking to superheat the wastewater sludge and convert it into small pelletized granules that are high in nitrogen and other organic materials. In New York City, for example, several sewage treatment plants have dewatering facilities that use large centrifuges along with the addition of chemicals such as polymer to further remove liquid from the sludge. The removed fluid, called centrate, is typically reintroduced into the wastewater process. The product which is left is called "cake" and that is picked up by companies which turn it into fertilizer pellets. This product is then sold to local farmers and turf farms as a soil amendment or fertilizer, reducing the amount of space required to dispose of sludge in landfills.

Treatment in the Receiving Environment

The outlet of a wastewater treating plant flows into a small river.

Many processes in a wastewater treatment plant are designed to mimic the natural treatment processes that occur in the environment, whether that environment is a natural water body or the ground. If not overloaded, bacteria in the environment will consume organic contaminants, although this will reduce the levels of oxygen in the water and may significantly change the overall ecology of the receiving water. Native bacterial populations feed on the organic contaminants, and the numbers of disease-causing microorganisms are reduced by natural environmental conditions such as predation exposure to ultraviolet radiation, for example. Consequently, in cases where the receiving environment provides a high level of dilution, a high degree of wastewater treatment may not be required. However, recent evidence has demonstrated that very low levels of certain contaminants in wastewater, including hormones (from animal husbandry and residue from human hormonal contraception methods) and synthetic materials such as phthalates that mimic hormones in their action, can have an unpredictable adverse impact on the natural biota and potentially on humans if the water is re-used for drinking water. In the United States and EU, uncontrolled discharges of wastewater to the environment are not permitted under law, and strict water quality requirements are to be met. A significant threat in the coming decades will be the increasing uncontrolled discharges of wastewater within rapidly developing countries.

Sewage Treatment in Developing Countries

There are few reliable figures on the share of the wastewater collected in sewers that is being treated in the world. In many developing countries the bulk of domestic and industrial wastewater is discharged without any treatment or after primary treatment only. In Latin America about 15 percent of collected wastewater passes through treatment plants (with varying levels of actual treatment). In Venezuela, a below average country in South America with respect to wastewater treatment, 97 percent of the country's sewage is discharged raw into the environment.

In a relatively developed Middle Eastern country such as Iran, Tehran's majority of population has totally untreated sewage injected to the city's groundwater. Israel has also aggressively pursued the use of treated sewer water for irrigation. In 2008, agriculture in Israel consumed 500 million cubic meters of potable water and an equal amount of treated sewer water. The country plans to provide a further 200 million cubic meters of recycled sewer water and build more desalination plants to supply even more water.

Most of sub-saharan Africa is without wastewater treatment.

Water utilities in developing countries are chronically underfunded because of low water tariffs, the nonexistence of sanitation tariffs in many cases, low billing efficiency (i.e. many users that are billed do not pay) and poor operational efficiency (i.e. there are overly high levels of staff, there are high physical losses, and many users have illegal connections and are thus not being billed). In addition, wastewater treatment typically is the process within the utility that receives the least attention, partly because enforcement of environmental standards is poor. As a result of all these factors, operation and maintenance of many wastewater treatment plants is poor. This is evidenced by the frequent breakdown of equipment, shutdown of electrically operated equipment due to power outages or to reduce costs, and sedimentation due to lack of sludge removal.

Developing countries as diverse as Egypt, Algeria, China or Colombia have invested substantial sums in wastewater treatment without achieving a significant impact in terms of environmental improvement. Even if wastewater treatment plants are properly operating, it can be argued that the environmental impact is limited in cases where the assimilative capacity of the receiving waters (ocean with strong currents or large rivers) is high, as it is often the case.

Enhanced Biological Phosphorus Removal

Enhanced Biological Phosphorus Removal is a process in which the biology within the activated sludge system is manipulated so that microorganisms take more phosphorus into their cells than they normally would.

It should be noted that all activated sludge systems involve biological phosphorus removal; phosphorus is an essential element for life. The key to EBPR is getting microorganisms to take in more phosphorus than they need.

This is also known as luxury uptake.

Phosphorous Removal from Wastewater

Controlling phosphorous discharged from municipal and industrial wastewater treatment plants is a key factor in preventing eutrophication of surface waters. Phosphorous is one of the major nutrients contributing in the increased eutrophication of lakes and natural waters. Its presence causes many water quality problems including increased purification costs, decreased recreational and conservation value of an impoundments, loss of livestock and the possible lethal effect of algal toxins on drinking water.

Phosphate removal is currently achieved largely by chemical precipitation, which is expensive and causes an increase of sludge volume by up to 40%. An alternative is the biological phosphate removal (BPR).

Phosphorous in Wastewater

Municipal wastewaters may contain from 5 to 20 mg/l of total phosphorous, of which 1-5 mg/l is organic and the rest in inorganic. The individual contribution tend to increase, because phosphorous is one of the main constituent of synthetic detergents. The individual phosphorous contribution varies between 0.65 and 4.80 g/inhabitant per day with an average of about 2.18 g. The usual forms of phosphorous found in aqueous solutions include:

- Orthophosphates: available for biological metabolism without further breakdown.
- Polyphosphates: molecules with 2 or more phosphorous atoms, oxygen and in some cases hydrogen atoms combine in a complex molecule. Usually polyphosphates undergo hydrolysis and revert to the orthophosphate forms. This process is usually quite slow.

Normally secondary treatment can only remove 1-2 mg/l, so a large excess of phosphorous is discharged in the final effluent, causing eutrophication in surface waters. New legislation requires a maximum concentration of P discharges into sensitive water of 2 mg/l.

Phosphorous Removal Processes

The removal of phosphorous from wastewater involves the incorporation of phosphate into TSS and the subsequent removal from these solids. Phosphorous can be incorporated into either biological solids (e.g. micro organisms) or chemical precipitates.

Phosphate Precipitation

Chemical precipitation is used to remove the inorganic forms of phosphate by the addition of a coagulant and a mixing of wastewater and coagulant. The multivalent metal ions most commonly used are calcium, aluminium and iron.

Calcium:

it is usually added in the form of lime $Ca(OH)_2$. It reacts with the natural alkalinity in the wastewater to produce calcium carbonate, which is primarily responsible for enhancing SS removal.

$$Ca(HCO_3)_2 + Ca(OH)_2 \text{ à } 2CaCO_3 \downarrow + 2H_2O$$

As the pH value of the wastewater increases beyond about 10, excess calcium ions will then react with the phosphate, to precipitate in hydroxylapatite:

$$10\ Ca^{2+} + 6\ PO_4^{3-} + 2\ OH^- \leftrightarrow Ca_{10}(PO_4)*6(OH)_2 \downarrow$$

Because the reaction is between the lime and the alkalinity of the wastewater, the quantity required will be, in general, independent of the amount of phosphate present. It will depend primarily on the alkalinity of the wastewater. The lime dose required can be approximated at 1.5 times the alkalinity as $CaCO_3$. Neutralisation may be required to reduce pH before subsequent treatment or disposal. Recarbonation with carbon dioxide (CO_2) is used to lower the pH value.

Aluminium and Iron:

Alum or hydrated aluminium sulphate is widely used precipitating phosphates and aluminium phosphates ($AlPO_4$). The basic reaction is:

$$Al^{3+} + H_nPO_4^{3-n} \leftrightarrow AlPO_4 + nH^+$$

This reaction is deceptively simple and must be considered in light of the many competing reactions and their associated equilibrium constants and the effects of alkalinity, pH, trace elements found in wastewater. The dosage rate required is a function of the phosphorous removal required. The efficiency of coagulation falls as the concentration of phosphorous decreases. In practice, an 80-90% removal rate is achieved at coagulant dosage rates between 50 and 200 mg/l. Dosages are generally established on the basis of bench-scale tests and occasionally by full-scale tests, especially if polymers are used. Aluminium coagulants can adversely affect the microbial population in activated sludge, especially protozoa and rotifers, at dosage rates higher than 150 mg/l. However this does not affect much either BOD or TSS removal, as the clarification function of protozoa and rotifers is largely compensated by the enhanced removal of SS by chemical precipitation.

Ferric chloride or sulphate and ferrous sulphate also know as copperas, are all widely used for phosphorous removal, although the actual reactions are not fully understood. The basic reaction is:

$$Fe^{3+} + H_nPO_4^{3-n} \leftrightarrow FePO_4 + nH^+$$

Ferric ions combine to form ferric phosphate. They react slowly with the natural alkalinity and so a coagulant aid, such as lime, is normally add to raise the pH in order to enhance the coagulation.

Strategies

The main phosphate removal processes are:

1. Treatment of raw/primary wastewater;
2. Treatment of final effluent of biological plants (postprecipitation);
3. Treatment contemporary to the secondary biologic reaction (co-precipitation).

The first process is included in the general category of chemical precipitation processes. Phosphorous is removed with 90% efficiency and the final P concentration is lower than 0.5 mg/l. The chemical dosage for P removal is the same as the dosage needed for BOD and SS removal, which uses the main part of these chemicals. As mentioned above lime consumption is dependent on the alkalinity of the wastewater: only 10% of the lime fed is used in the phosphorous removal reaction. The remaining amount reacts with water alkalinity, with softening. To determine the lime quantity needed it is possible to use diagrams: i.e. the lime used to reach ph 11 is 2-2.5 times water alkalinity.

The postprecipitation is a standard treatment of a secondary effluent, usually using only metallic reagents. It is the process that gives the highest efficiency in phosphorous removal. Efficiency can reach 95%, and P concentration in the effluent can be lower than 0.5 mg/l. Postprecipitation gives also a good removal of the SS that escape the final sedimentation of the secondary process.

Its advantage is also to guarantee purification efficiency at a certain extent even if the biological process is not efficient for some reason. The chemical action is stronger, since the previous biologic treatment transforms part of the organic phosphates in orthophosphates. Disadvantages are high costs for the treatment plant (big ponds and mixing devices) and sometimes a too dilute effluent. Using ferric salts there is also the risk of having some iron in the effluent, with residual coloration. The metallic ions dosage is about 1.5-2.5 ions for every phosphorus ion (on average about 10-30 g/mc of water).

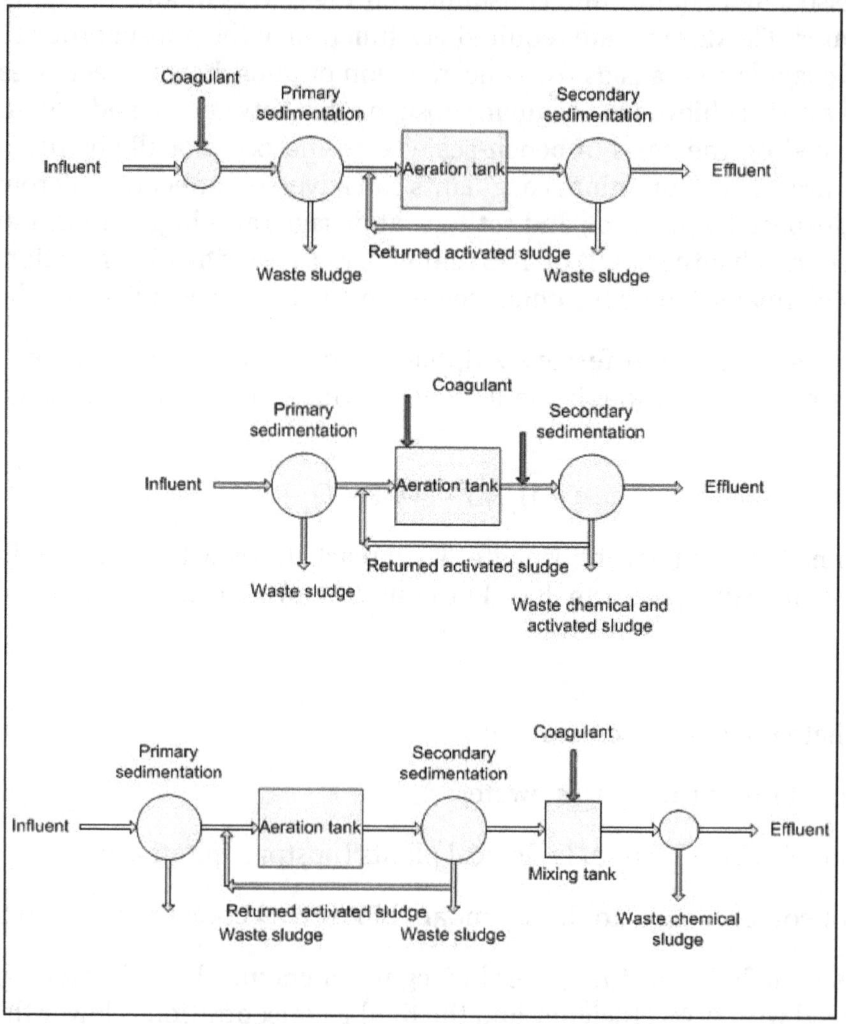

The coprecipitation process is particularly suitable for active sludge plants, where the chemicals are fed directly in the aeration tank or before it. The continuous sludge recirculation, together with the coagulation-flocculation and adsorption process due to active sludge, allows a reduction in chemical consumption. Moreover the costs for the plant are lower, since there is no need for big postprecipitation ponds. In this process the chemical added are only iron and aluminium, lime is added only for pH correction. Lower costs and more simplicity are contrasted by a phosphorous removal efficiency lower than with postprecipitation (below 85%). The phosphorous concentration in the final effluent is about 1 mg/l. Another disadvantage is that biological and chemical sludge are mixed, so they cannot be used separately in next stages. Mixed sludges need bigger sedimentation tanks than activated sludge.

Biological Processes

Over the past 20 years, several biological suspended growth process configurations have been used to accomplish biological phosphorous removal. The most important are shown in the following picture.

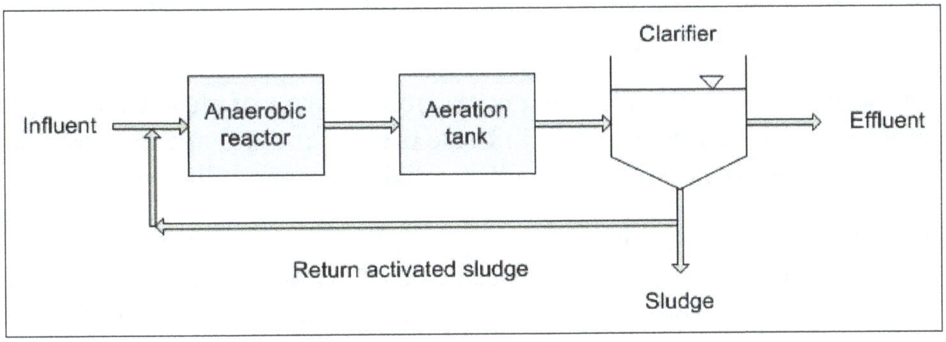

The principal advantages of biological phosphorous removal are reduced chemical costs and less sludge production as compared to chemical precipitation.

In the biological removal of phosphorous, the phosphorous in the influent wastewater is incorporated into cell biomass, which is subsequently removed from the process as a result of sludge wasting. The reactor configuration provides the P accumulating organisms (PAO) with a competitive advantage over other bacteria. So PAO are encouraged to grow and consume phosphorous. The reactor configuration in comprised of an anaerobic tank and an activated sludge activated tank. The retention time in the anaerobic tank is about 0.50 to 1.00 hours and its contents are mixed to provide contact with the return activated sludge and influent wastewater.

In the anaerobic zone: Under anaerobic conditions, PAO assimilate fermentation products (i.e. volatile fatty acids) into storage products within the cells with the concomitant release of phosphorous from stored polyphosphates. Acetate is produced by fermentation of bsCOD, which is dissolved degradable organic material that can be easily assimilated by the biomass. Using energy available from stored polyphosphates, the PAO assimilate acetate and produce intracellular polyhydroxybutyrate (PHB) storage products. Concurrent with the acetate uptake is the release of orthophosphates, as well as magnesium, potassium, calcium cations. The PHB content in the PAO increases as the polyphosphate decreases.

In the aerobic zone: energy is produced by the oxidation of storage products and polyphosphate storage within the cell increases. Stored PHB is metabolized, providing energy from oxidation and carbon for new cell growth. Some glycogen is produced from PHB metabolism. The energy released from PHB oxidation is used to form polyphosphate bonds in cell storage. The soluble orthophosphate is removed from solution and incorporated into polyphosphates within the bacterial cell. PHB utilisation also enhances cell growth and this new biomass with high polyphosphate storage accounts for phosphorous removal. As a portion of the biomass is wasted, the stored phosphorous is removed from the biotreatment reactor for ultimate disposal with the waste sludge.

The amount of phosphorous removed by biological storage can be estimated from the amount of bsCOD that is available in the wastewater influent. Better performance for BPR systems is achieved when bsCOD acetate is available at a steady rate.

Ultraviolet Germicidal Irradiation

Ultraviolet germicidal irradiation (UVGI) is a sterilization method that uses ultraviolet (UV) light at sufficiently short wavelength to break down micro-organisms. It is used in a variety of applications, such as food, air and water purification. UV has been a known mutagen at the cellular level for more than 100 years. The 1903 Nobel Prize for Medicine was awarded to Niels Finsen for his use of UV against tuberculosis.

UVGI utilises the short wavelength of UV that is harmful to forms of life at the micro-organic level. It is effective in destroying the nucleic acids in these organisms so that their DNA is disrupted by the UV radiation. This removes their reproductive capabilities and kills them.

The wavelength of UV that causes this effect is rare on Earth as its atmosphere blocks it. Using a UVGI device in certain environments like circulating air or water systems creates a deadly effect on micro-organisms such as pathogens, viruses and moulds that are in these environments. Coupled with a filtration system, UVGI can remove harmful micro-organisms from these environments.

The application of UVGI to sterilization has been an accepted practice since the mid-20th century. It has been used primarily in medical sanitation and sterile work facilities. Increasingly it was employed to sterilize drinking and wastewater, as the holding facilities were enclosed and could be circulated to ensure a higher exposure to the UV. In recent years UVGI has found renewed application in air sanitisation.

Method of Operation

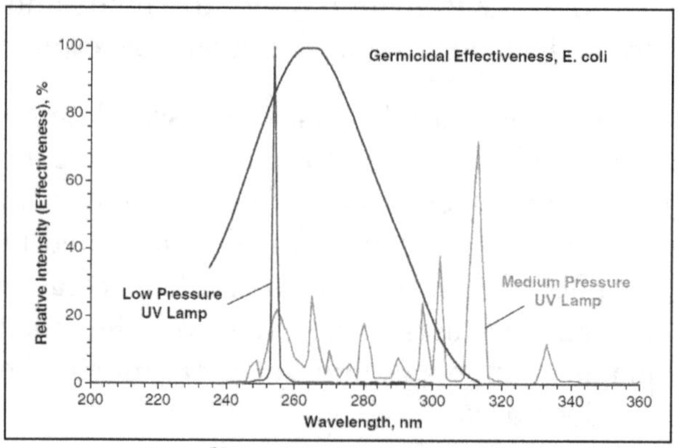

Low-pressure & medium-pressure mercury lamps compared to e.coli germicidal effectiveness curve. Ultraviolet Germicidal Irradiation Handbook.

UV light is electromagnetic radiation with wavelengths shorter than visible light. UV can be separated into various ranges, with short-wavelength UV (UVC) considered "germicidal UV". At certain wavelengths, UV is mutagenic to bacteria, viruses and other microorganisms. Particularly at wavelengths around 260 nm–270 nm, UV breaks molecular bonds within microorganismal DNA, producing thymine dimers that can kill or disable the organisms.

- Mercury-based lamps emit UV light at the 253.7 nm line.
- Ultraviolet Light Emitting Diodes (UV-C LED) lamps emit UV light at selectable wavelengths between 255 and 280 nm.

- Pulsed-xenon lamps emit UV light across the entire UV spectrum with a peak emission near 230nm.

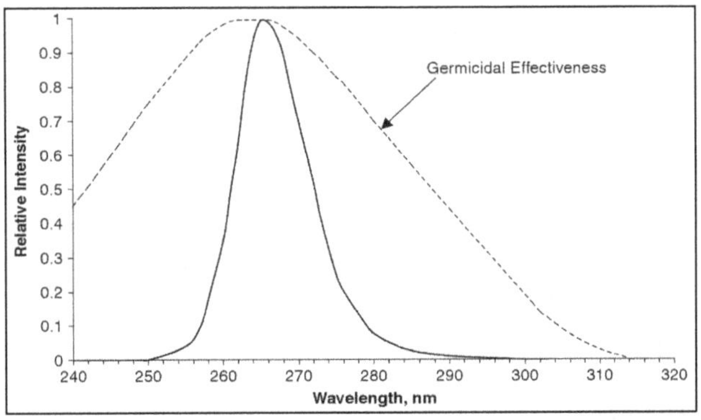

UV-C LED emitting 265 nm compared to e.coli germicidal effectiveness curve. Ultraviolet Germicidal Irradiation Handbook.

This process is similar to the effect of longer wavelengths (UVB) producing sunburn in humans. Microorganisms have less protection against UV, and cannot survive prolonged exposure to it.

A UVGI system is designed to expose environments such as water tanks, sealed rooms and forced air systems to germicidal UV. Exposure comes from germicidal lamps that emit germicidal UV at the correct wavelength, thus irradiating the environment. The forced flow of air or water through this environment ensures exposure.

Effectiveness

The effectiveness of germicidal UV depends on the length of time a microorganism is exposed to UV, the intensity and wavelength of the UV radiation, the presence of particles that can protect the microorganisms from UV, and a microorganism's ability to withstand UV during its exposure.

In many systems, redundancy in exposing microorganisms to UV is achieved by circulating the air or water repeatedly. This ensures multiple passes so that the UV is effective against the highest number of microorganisms and will irradiate resistant microorganisms more than once to break them down.

"Sterilization" is often misquoted as being achievable. While it is theoretically possible in a controlled environment, it is very difficult to prove and the term "disinfection" is generally used by companies offering this service as to avoid legal reprimand. Specialist companies will often advertise a certain log reduction e.g., 99.9999% effective, instead of sterilization. This takes into consideration a phenomenon known as light and dark repair (photoreactivation and base excision repair, respectively), in which a cell can repair DNA that has been damaged by UV light.

The effectiveness of this form of disinfection depends on line-of-sight exposure of the microorganisms to the UV light. Environments where design creates obstacles that block the UV light are not as effective. In such an environment, the effectiveness is then reliant on the placement of the UVGI system so that line of sight is optimum for disinfection.

Dust and films coating the bulb lower UV output. Therefore, bulbs require periodic cleaning and replacement to ensure effectiveness. The lifetime of germicidal UV bulbs varies depending on design. Also, the material that the bulb is made of can absorb some of the germicidal rays.

Lamp cooling under airflow can also lower UV output; thus, care should be taken to shield lamps from direct airflow, or to add additional lamps to compensate for the cooling effect.

Increases in effectiveness and UV intensity can be achieved by using reflection. Aluminum has the highest reflectivity rate versus other metals and is recommended when using UV.

One method for gauging UV effectiveness in water disinfection applications is to compute UV dose. The U.S. EPA publishes UV dosage guidelines for water treatment applications. UV dose cannot be measured directly but can be inferred based on the known or estimated inputs to the process:

- Flow rate (contact time),
- Transmittance (light reaching the target),
- Turbidity (cloudiness),
- Lamp age or fouling or outages (reduction in UV intensity).

In air and surface disinfection applications the UV effectiveness is estimated by calculating the UV dose which will be delivered to the microbial population. The UV dose is calculated as follows:

UV dose $\mu Ws/cm^2$ = UV intensity $\mu W/cm^2$ x Exposure time (seconds).

The UV intensity is specified for each lamp at a distance of 1 meter. UV intensity is inversely proportional to the square of the distance so it decreases at longer distances. Alternatively, it rapidly increases at distances shorter than 1m. In the above formula the UV intensity must always be adjusted for distance unless the UV dose is calculated at exactly 1m from the lamp. Also, to ensure effectiveness the UV dose must be calculated at the end of lamp life (EOL is specified in number of hours when the lamp is expected to reach 80% of its initial UV output) and at the furthest distance from the lamp on the periphery of the target area. Some *shatter-proof* lamps are coated with a fluorated ethylene polymer to contain glass shards and mercury in case of breakage; this coating reduces UV output by as much as 20%.

To accurately predict what UV dose will be delivered to the target the UV intensity, adjusted for distance, coating and end of lamp life, will be multiplied by the exposure time. In static applications the exposure time can be as long as needed for an effective UV dose to be reached. In case of rapidly moving air, in AC air ducts for example, the exposure time is short so the UV intensity must be increased by introducing multiple UV lamps or even banks of lamps. Also, the UV installation must be located in a long straight duct section with the lamps perpendicular to the air flow to maximize the exposure time.

These calculations actually predict the UV fluence and it is assumed that the UV fluence will be equal to the UV dose. The UV dose is the amount of germicidal UV energy absorbed by a microbial population over a period of time. If the microorganisms are planktonic (free floating) the UV fluence will be equal the UV dose. However, if the microorganisms are protected by mechanical particles, such as dust and dirt, or have formed biofilm a much higher UV fluence will be needed for an effective UV dose to be introduced to the microbial population.

Inactivation of Microorganisms

The degree of inactivation by ultraviolet radiation is directly related to the UV dose applied to the water. The dosage, a product of UV light intensity and exposure time, is usually measured in microjoules per square centimeter, or equivalently as microwatt seconds per square centimeter ($\mu W{\cdot}s/cm^2$). Dosages for a 90% kill of most bacteria and viruses range from 2,000 to 8,000 $\mu W{\cdot}s/cm^2$. Larger parasites such as cryptosporidium require a lower dose for inactivation. As a result, the U.S. Environmental Protection Agency has accepted UV disinfection as a method for drinking water plants to obtain cryptosporidium, giardia or virus inactivation credits. For example, for one-decimal-logarithm reduction of cryptosporidium, a minimum dose of 2,500 $\mu W{\cdot}s/cm^2$ is required based on the U.S. EPA UV Guidance Manual published in 2006.

Creating UVGI

Germicidal UV is delivered by a mercury-vapor lamp that emits UV at the germicidal wavelength. Mercury vapour emits at 254nm. Many germicidal UV bulbs use special transformers to ensure even electrical flow to the bulbs so the correct wavelength is maintained. Since germicidal UV has a narrow bandwidth, power fluctuations will render intended irradiating environments ineffective. In some cases, UVGI electrodeless lamps can be energised with microwaves, giving very long stable life and other advantages. This is known as 'Microwave UV.'

There are several different types of germicidal lamps: - Low-pressure UV lamps offer high efficiencies (approx 35% UVC) but lower power, typically 1 W/cm^3 power density. - Amalgam UV lamps are a high-power version of low-pressure lamps. They operate at higher temperatures and have a lifetime of up to 16,000 hours. Their efficiency is slightly lower than that of traditional low-pressure lamps (approx 33% UVC output) and power density is approx 2-3 W/cm^3. - Medium-pressure UV lamps have a broad and pronounced peak-line spectrum and a high radiation output but lower UVC efficicncy of 10% or less. Typical power density is 30 W/cm^3 or greater.

Depending on the quartz glass used for the lamp body, low-pressure and amalgam UV lamps emit light at 254 nm and 185 nm (for oxidation).

185 nm light is used to generate ozone.

Strengths and Weaknesses

Advantages

UV water treatment devices can be used for well water and surface water disinfection. UV treatment compares favorably with other water disinfection systems in terms of cost, labor, and the need for technically trained personnel for operation. Water chlorination treats larger organisms and offers residual disinfection, but these systems are expensive because they need special operator training and a steady supply of a potentially hazardous material. Finally, boiling of water is the most reliable treatment method but it demands labor, and imposes a high economic cost. UV treatment is rapid and, in terms of primary energy use, approximately 20,000 times more efficient than boiling.

Disadvantages

UV disinfection is most effective for treating high-clarity, purified reverse osmosis distilled water. Suspended particles are a problem because microorganisms buried within particles are shielded from the UV light and pass through the unit unaffected. However, UV systems can be coupled with a pre-filter to remove those larger organisms that would otherwise pass through the UV system unaffected. The pre-filter also clarifies the water to improve light transmittance and therefore UV dose throughout the entire water column. Another key factor of UV water treatment is the flow rate—if the flow is too high, water will pass through without sufficient UV exposure. If the flow is too low, heat may build up and damage the UV lamp.

A disadvantage of UVGI is that while water treated by chlorination is resistant to reinfection (until the chlorine off-gasses), UVGI water is not resistant to reinfection. UVGI water must be transported or delivered in such a way as to avoid reinfection.

Technology

Lamps

A 9 W germicidal lamp in a compact fluorescent lamp form factor.

Germicidal UV for disinfection is most typically generated by a mercury-vapor lamp. Low-pressure mercury vapor has a strong emission line at 254 nm, which is within the range of wavelengths that demonstrate strong disinfection effect. The optimal wavelengths for disinfection are close to 270 nm.

Lamps are either amalgam or medium-pressure lamps. Low-pressure UV lamps offer high efficiencies (approx 35% UVC) but lower power, typically 1 W/cm power density (power per unit of arc length). Amalgam UV lamps are a higher-power version of low-pressure lamps. They operate at higher temperatures and have a lifetime of up to 16,000 hours. Their efficiency is slightly lower than that of traditional low-pressure lamps (approx 33% UVC output) and power density is approximately 2–3 W/cm. Medium-pressure UV lamps have a broad and pronounced peak-line spectrum and a high radiation output but lower UVC efficiency of 10% or less. Typical power density is 30 W/cm^3 or greater.

Depending on the quartz glass used for the lamp body, low-pressure and amalgam UV emit radiation at 254 nm and also at 185 nm, which has chemical effects. UV radiation at 185 nm is used to generate ozone.

The UV lamps for water treatment consist of specialized low-pressure mercury-vapor lamps that produce ultraviolet radiation at 254 nm, or medium-pressure UV lamps that produce a polychromatic output from 200 nm to visible and infrared energy. The UV lamp never contacts the water; it is either housed in a quartz glass sleeve inside the water chamber or mounted external to the water which flows through the transparent UV tube. Water passing through the flow chamber is exposed to UV rays which are absorbed by suspended solids, such as microorganisms and dirt, in the stream.

Compact and versatile options with UV-C LEDs.

Light Emitting Diodes (LEDs)

Recent developments in LED technology have led to commercially available UV-C LEDs. UV-C LEDs use semiconductors to emit light between 255 nm-280 nm. The wavelength emission is tuneable by adjusting the material of the semiconductor. The reduced size of LEDs opens up options for small reactor systems allowing for point-of-use applications and integration into medical devices. Low power consumption of semiconductors introduce UV disinfection systems that utilized small solar cells in remote or Third World applications.

Water Treatment Systems

Sizing of a UV system is affected by three variables: flow rate, lamp power, and UV transmittance in the water. Manufacturers typically developed sophisticated Computational Fluid Dynamics (CFD) models validated with bioassay testing. This involves testing the UV reactor's disinfection performance with either MS2 or T1 bacteriophages at various flow rates, UV transmittance, and power levels in order to develop a regression model for system sizing. For example, this is a requirement for all drinking water systems in the United States per the EPA UV Guidance Manual.[:5-2]

The flow profile is produced from the chamber geometry, flow rate, and particular turbulence model selected. The radiation profile is developed from inputs such as water quality, lamp type (power, germicidal efficiency, spectral output, arc length), and the transmittance and dimension of the quartz sleeve. Proprietary CFD software simulates both the flow and radiation profiles. Once the 3D model of the chamber is built, it is populated with a grid or mesh that comprises thousands of small cubes.

Points of interest—such as at a bend, on the quartz sleeve surface, or around the wiper mechanism—use a higher resolution mesh, whilst other areas within the reactor use a coarse mesh. Once the mesh is produced, hundreds of thousands of virtual particles are "fired" through the chamber. Each particle has several variables of interest associated with it, and the particles are "harvested" after the reactor. Discrete phase modeling produces delivered dose, head loss, and other chamber-specific parameters.

When the modeling phase is complete, selected systems are validated using a professional third party to provide oversight and to determine how closely the model is able to predict the reality of system performance. System validation uses non-pathogenic surrogates such as MS 2 phage or *Bacillus subtilis* to determine the Reduction Equivalent Dose (RED) ability of the reactors. Most systems are validated to deliver 40 mJ/cm^2 within an envelope of flow and transmittance.

To validate effectiveness in drinking-water systems, the method described in the EPA UV Guidance Manual is typically used by the U.S., whilst Europe has adopted Germany's DVGW 294 standard. For wastewater systems, the NWRI/AwwaRF Ultraviolet Disinfection Guidelines for Drinking Water and Water Reuse protocols are typically used, especially in wastewater reuse applications.

Potential Dangers

At certain wavelengths (including UVC) UV is harmful to humans and other forms of life. In most UVGI systems the lamps are shielded or are in environments that limit exposure, such as a closed water tank or closed air circulation system, often with interlocks that automatically shut off the UV lamps if the system is opened for access by human beings. Limited exposure mitigates the risk of danger.

In human beings, skin exposure to germicidal wavelengths of UV light can produce sunburn and (in some cases) skin cancer. Exposure of the eyes to this UV radiation can produce extremely painful inflammation of the cornea and temporary or permanent vision impairment, up to and including blindness in some cases. UV can damage the retina of the eye.

Another potential danger is the UV production of ozone. UVC light from the sun is partly responsible for the earth's ozone layer in the stratosphere, but ozone at the atmospheric level can be harmful to a person's health. The United States Environmental Protection Agency designated .05 parts per million (ppm) of ozone to be a safe level.

UV-C radiation is able to break-down chemical bonds. This leads to rapid ageing of plastics (insulations, gasket) and other materials. Note that plastics sold to be "UV-resistant" are tested only for UV-B, as UV-C doesn't normally reach the surface of the Earth.

Uses for UVGI

Air Purification

UVGI can be used to sterilize air that passes UV lamps via forced air. Air purification UVGI systems can be freestanding units with shielded UV lamps that use a fan to force air past the UV light. Other systems are installed in forced air systems so that the circulation for the premises moves micro-organisms past the lamps. Key to this form of sterilization is placement of the UV lamps and a good filtration system to remove the dead micro-organisms. For example, forced air systems by design impede line of sight, thus creating areas of the environment that will be shaded from the UV light. However, a UV lamp placed at the coils and drain pan of cooling system will keep micro-organisms from forming in these naturally damp places. The most effective method for treating the air rather than the coils is in-line duct systems, these systems are placed in the center of the duct and parallel to the air flow.

Water Purification

Water purification via UVGI is used in most water sterilization processes, such as purification, detoxification and disinfection. Its use in wastewater treatment is replacing chlorination due to that chemical's toxic by-products. A disadvantage is that water treated by chlorination is resistant to reinfection, where UVGI water must be transported and delivered in such a way as to avoid contamination. Individual waste streams to be treated by UVGI must be tested to ensure that the method will be effective due to potential interferences such as suspended solids, dyes or other substances that may block or absorb the UV radiation.

"UV units to treat small batches (1 to several liters) or low flows (1 to several liters per minute) of water at the community level are estimated to have costs of 0.02 US$ per 1000 liters of water, including the cost of electricity and consumables and the annualized capital cost of the unit." (WHO).

Laboratory Hygiene

UVGI is often used to sterilize equipment such as safety goggles, instruments, pipettors, and other devices. Lab personnel also sterilize glassware and plasticware this way. Microbiology laboratories use UVGI to sterilize surfaces inside biological safety cabinets ("hoods") between uses.

Septic Tank

A septic tank collects and treats wastewater at a property that is not connected to the mains sewer system.

Installed underground, a septic tank makes use of natural processes to treat the sewage it stores. Usually made up of two chambers or compartments, the tank receives wastewater from an inlet pipe.

Wastewater enters the first chamber and separates over time, with solids settling at the bottom, oils and greases forming a layer of scum at the top, and a layer of relatively clear water remaining in the middle.

This clarified wastewater enters the second chamber. It then exits via an outlet pipe into a septic drain field, also known as a seepage field or leach field. The remaining scum and solids in the tank are broken down by naturally occurring bacteria and what is left should be professionally removed periodically.

The septic tank was patented in London around 1900. Websters Dictionary defines the septic tank as "A tank in which waste matter is decomposed through bacterial action." The modern septic tank is a watertight box usually made of precast concrete, concrete blocks, or reinforced fiberglass.

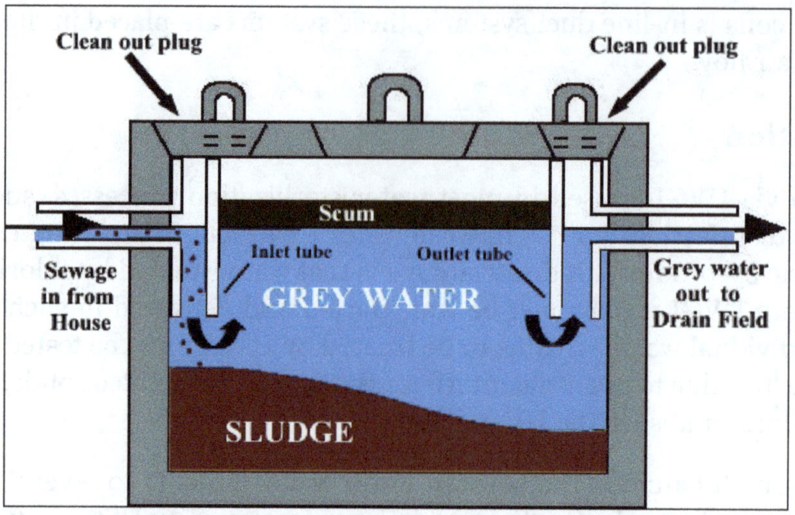

The septic system is a small, on-site treatment and disposal system buried in the ground. the septic system has two essential parts: (1) the septic tank and (2) the soil absorption area. When household waste enters the septic tank several things occur:

1. Organic solid material floats to the surface and forms a layer of what is commonly called "scum." Bacteria in the septic tank biologically convert this material to liquid.
2. Inorganic or inert solid materials and the by-products of bacterial digestions sink to the bottom of the tank and form a layer commonly known as "sludge."
3. Only clear water should exist between the scum and sludge layers. It is this clear water – And only this clear water – that should overflow into the soil absorption area.

Septic Tank with Baffle

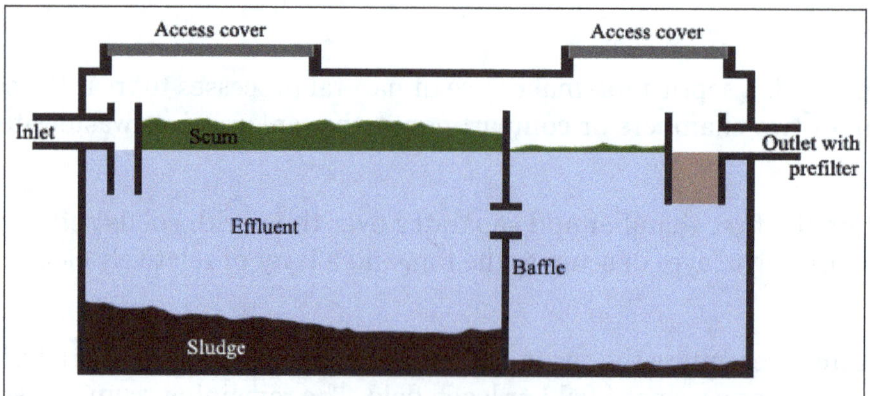

Fosse Septique

Solid material overflowing into the soil absorption area should be avoided at all costs. It is this solids overflow that clogs soil pores and causes system to fail. two main factors cause solid material to build up enough to overflow: (1) bacterial deficency, and (2) lack of sludge removal.

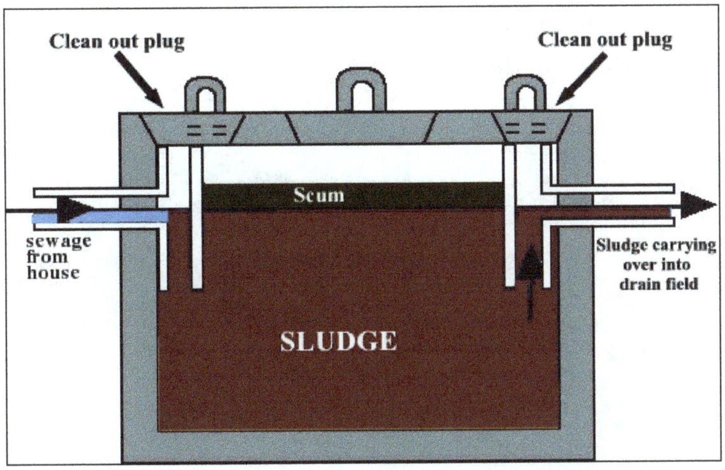

Bacteria must be present in the septic tank to break down and digest the organic solids. Normal household waste probides enough bacteria to digest the solids unless any harm is done to the bacteria. Bacteria are very sensitive to environmental changes. Check teh lables of products you normally use in home. Products carrying harsh warnings such as "Harmful or Fatal if Swallowed" will harm bacteria.

- Detergents,
- Bleaches,
- Cleaning compounds,
- Disinfectants,
- Acids,
- Toilet cleaners,
- Polishes,
- Caustic drain openers.

People rarely think of the effect of these products on the septic tank system when the products go down the drain. What kind of effect to you think anti-septics have on your septic tank?

Bacteria must be present to digest the scum. If not digested, the scum will accumulate untill it overflows, clogging the soil absorption area.

The sludge in the septic tank – inorganic and inert material – is not biodegradable and will not decompose. If not removed, the sludge will accumulate until it eventually overflows, again clogging the soil absorption area.

Emptying

Waste that is not decomposed by the anaerobic digestion must eventually be removed from the septic tank. Otherwise the septic tank fills up and wastewater containing undecomposed material discharges directly to the drainage field. Not only is this detrimental for the environment but, if the sludge overflows the septic tank into the leach field, it may clog the leach field piping or decrease the soil porosity itself, requiring expensive repairs.

A vacuum truck used to empty septic tanks in Germany.

When a septic tank is emptied, the accumulated sludge (septage, also known as fecal sludge) is pumped out of the tank by a vacuum truck. How often the septic tank must be emptied depends on the volume of the tank relative to the input of solids, the amount of indigestible solids, and the ambient temperature (because anaerobic digestion occurs more efficiently at higher temperatures), as well as usage, system characteristics and the requirements of the relevant authority. Some health authorities require tanks to be emptied at prescribed intervals, while others leave it up to the decision of an inspector. Some systems require pumping every few years or sooner, while others may be able to go 10–20 years between pumpings. An older system with an undersize tank that is being used by a large family will require much more frequent pumping than a new system used by only a few people. Anaerobic decomposition is rapidly restarted when the tank is refilled.

Services for de-sludging tend to empty a septic tank completely, i.e. take out all septage, while the actual requirement is removal of settled solids, and even this purposefully incompletely so as to leave at least some of the microbial populations in place to continue the anaerobic degradation processes that take place in a septic tank.

Maintenance

A septic tank before installation, with manhole cover on top.

The same tank partially installed in the ground.

Like any system, a septic system requires maintenance. The maintenance of a septic system is often the responsibility of the resident or property owner. Some forms of abuse or neglect include the following:

User's Actions

- Excessive disposal of cooking oils and grease can cause the inlet drains to block. Oils and grease are often difficult to degrade and can cause odor problems and difficulties with the periodic emptying.
- Flushing non-biodegradable waste items down the toilet such as cigarette butts, cotton buds/swabs or menstrual hygiene products and condoms can cause a septic tank to clog and fill rapidly, so these materials should not be disposed of in that manner. The same applies when the toilet is connected to a sewer rather than a septic tank.
- Using the toilet for disposal of food waste can cause a rapid overload of the system with solids and contribute to failure.
- Certain chemicals may damage the components of a septic tank or kill the bacteria needed in the septic tank for the system to operate properly, such as pesticides, herbicides, materials with high concentrations of bleach or caustic soda (lye), or any other inorganic materials such as paints or solvents.
- Using water softeners - the brine discharge from water softeners may harm the bacteria responsible for breaking down the wastewater. Usually the brine is however sufficiently diluted with other wastewater that it does not adversely affect the septic system.

Other Factors

- Roots from trees and shrubbery protruding above the tank or drainfield may clog and/or rupture them. Trees that are directly within the vicinity of a concrete septic tank have the potential to penetrate the tank as the system ages and the concrete begins to develop cracks and small leaks. Tree roots can cause serious flow problems due to plugging and blockage of drain pipes, added to which the trees themselves tend to expand extremely vigorously due to the ready supply of nutrients from the septic system.

- Playgrounds and storage buildings may cause damage to a tank and the drainage field. In addition, covering the drainage field with an impermeable surface, such as a driveway or parking area, will seriously affect its efficiency and possibly damage the tank and absorption system.
- Excessive water entering the system may overload it and cause it to fail.
- Very high rainfall, rapid snowmelt, and flooding from rivers or the sea can all prevent a drain field from operating, and can cause flow to back up, interfering with the normal operation of the tank. High winter water tables can also result in groundwater flowing back into the septic tank.
- Over time, biofilms develop on the pipes of the drainage field, which can lead to blockage. Such a failure can be referred to as "biomat failure".

Septic Tank Additives

Septic tank additives have been promoted by some manufacturers with the aim to improve the effluent quality from septic tanks, reduce sludge build-up and to reduce odors. However, these additives—which are commonly based on "effective microorganisms"—are usually costly in the longer term and fail to live up to expectations. It has been estimated that in the U.S. more than 1,200 septic system additives were available on the market in 2011. However, very little peer-reviewed and replicated field research exists regarding the efficacy of these biological septic tank additives.

Environmental Concerns

While a properly maintained and located septic tank does not pose any more environmental problems than centralized municipal sewage treatment, certain problems can arise with septic tanks in unsuitable locations. Since septic systems require large drainfields, they are not suitable for densely built areas.

Odor and Gas Emissions

Some constituents of wastewater, especially sulfates, under the anaerobic conditions of septic tanks, are reduced to hydrogen sulfide, a pungent and toxic gas. Methane may also be released. Nitrates and organic nitrogen compounds can be reduced to ammonia. Because of the anaerobic conditions, fermentation processes take place, which may generate carbon dioxide and/or methane.

Nutrients in the Effluent

Septic tanks by themselves are ineffective at removing nitrogen compounds that have potential to cause algal blooms in waterways into which affected water from a septic system finds its way. This can be remedied by using a nitrogen-reducing technology, or by simply ensuring that the leach field is properly sited to prevent direct entry of effluent into bodies of water.

The fermentation processes cause the contents of a septic tank to be anaerobic with a low redox potential, which keeps phosphates in a soluble and, thus, mobilized form. Phosphates discharged from a septic tank into the environment can trigger prolific plant growth including algal blooms, which can also include blooms of potentially toxic cyanobacteria.

The soil's capacity to retain phosphorus is usually large enough to handle the load through a normal residential septic tank. An exception occurs when septic drain fields are located in sandy or coarser soils on property adjoining a water body. Because of limited particle surface area, these soils can become saturated with phosphates. Phosphates will progress beyond the treatment area, posing a threat of eutrophication to surface waters.

Groundwater Pollution

In areas with high population density, groundwater pollution beyond acceptable limits may occur. Some small towns are experiencing the costs of building very expensive centralized wastewater treatment systems because of this problem, owing to the high cost of extended collection systems. To reduce residential development which might increase the demand to construct an expensive centralized sewerage system, building moratoriums and limits on the subdivision of property are often imposed. Ensuring existing septic tanks are functioning properly can also be helpful for a limited time, but becomes less effective as a primary remediation strategy as population density increases.

Surface Water Pollution

In areas adjacent to water bodies with fish or shellfish intended for human consumption, improperly maintained and failing septic systems contribute to pollution levels that can force harvest restrictions and/or commercial or recreational harvest closures.

Septic Tank Pumping

A large majority of on-lot sewage systems have septic tanks. The question of how often a septic tank should be pumped has been debated for many years. On the one hand you will find homeowners who claim they have never pumped their septic tank and it "seems" to work perfectly. On the other hand, in an attempt to create a uniform pumping policy, regulators have come down on the conservative side and have stated that all septic tanks should be pumped every two or three years.

How Frequent should a Septic Tank be Pumped?

The frequency of pumping depends on several factors:

- Capacity of septic tank.
- Daily volume of wastewater added to the septic tank.
- Amount of solids in wastewater stream. It should be noted that there are several classes of solids that are commonly put into a septic tank. These include (1) biodegradable "organic" solids such as feces, (2) slowly biodegradable "organic" solids such as toilet paper and cellulosic compounds, which take a long time to biodegrade in the septic tank, and (3) non-biodegradable solids such as kitty litter, plastics, etc., which do not biodegrade and quickly fill the septic tank. Reducing the amount of slowly biodegradable organics and

non-biodegradable waste added to your septic tank will greatly reduce the rate at which solids accumulate in the tank.

Another contributor to how quickly a septic tank will fill with solids is life style. The two most important life-style issues related to septic tank performance are:

1. Water usage in the home,
2. Age of the residents.

Homes with growing families including children ranging from small children to teenagers usually use more water and put more solids into the septic tank. On the other hand, empty nesters, and especially the elderly tend to use much less water and put smaller amounts of solids into septic tanks.

Another important consideration regarding how often a septic tank should be pumped is timing. As stated earlier, as a septic tank fills with solids, these solids tend to be carried from the tank to the soil absorption area, especially from tanks that do not have exit filters. As more solids accumulate in the absorption area, these solids begin to clog the soil and restrict the movement of wastewater into the soil. By the time sewage has backed up into the home, the soil absorption area is clogged with a nearly impermeable biomat and flooded with wastewater because the soil is no longer able to absorb the wastewater produced on a daily basis. Removal of these biomats is usually expensive and time consuming. Pumping the septic tank will not remove the biomat. Removing the biomat requires that you pump the wastewater ponded in the soil absorption area. Then the pump-access hole to the absorption area should be left open for several days. Once the absorption area is free of water and has become aerated, the biomat usually decomposes in a few days. When the absorption area is pumped, the septic tank should also be pumped, thus enhancing the development of the aerated conditions in the absorption area.

Time to Pump your Septic Tank

So, how does one decide how often a septic tank should be pumped? We know homes that put large amounts of non-biodegradable and slowly biodegradable organics into the septic tank need to pump more often. We also know that the septic tank should be pumped before the captured solids accumulate to the point where these solids begin being carried with the tank effluent to the absorption area. There are two relatively safe approaches to deciding when (or how often) to pump your septic tank. One is to just have it pumped every two or three years. The other is to open the access port to the first chamber once every year and insert a long pole to the bottom of the tank and withdraw it. You can see the depth of sludge by the darkness on the pole. If the sludge is more than a third of the tank depth, it is time to have it pumped. Most homeowners are better off just having their tank pumped every two or three years.

The Pumping Process

Septic tank pumping and haul contractors can pump your septic tank. It is a good idea to be on hand to ensure that it is done properly. To extract all the material from the tank, the scum layer must be broken up and the sludge layer mixed with the liquid portion of the tank. This is usually done by alternately pumping liquid from the tank and re-injecting it into the bottom of the tank. The septic tank must be pumped through the two large central access ports (manholes), not the small inlet or outlet inspection ports located above each baffle. Pumping a tank through the baffle inspection ports can damage the baffles and yield incomplete removal of sludge and scum.

The use of additives in septic tanks to reduce the sludge volume or as a substitute for pumping is not recommended. In fact, relying on additives rather than conventional tank pumping may result in failure of the entire septic system.

When you have your septic tank pumped, an additional step may help keep your septic system functioning properly for a long time. Most companies that pump septic tanks also have a certified PSMA Inspector in their company. This inspector can tell you if your septic tank needs repair or if other components of your septic system need maintenance.

To facilitate future cleaning and inspection, install risers from the central access ports and inspection ports to the soil surface. Also mark the location of the tank, so it can be easily located for future pumping.

Schedule Septic Tank Pumping

Homeowners should get in the habit of having the septic tank pumped. If you are able and willing to have your septic tank pumped on a routine basis (such as every two or three years), it may be possible to further enhance the effectiveness of your entire on-lot wastewater disposal system. Research at Penn State has shown that your soil absorption system will benefit from periodic resting (a period during which no wastewater is added to the absorption area). To get the greatest benefit from pumping your septic tank, it is recommended that you have your septic tank pumped every two to three years on the day before you, and your family, leave for your summer vacation. This means the whole system, especially the soil absorption area, will have the opportunity to dry out and any partially decomposed organic waste (biomat) that may have developed in the soil absorption area will quickly decompose in the absence of water.

Aerobic Treatment Unit

An aerobic treatment unit (ATU) consists of several processes that function together to provide a high quality effluent. These are gross solids (trash) removal, aeration, clarification, and sludge return. These processes are generally contained within separate chambers of a single tank. A series of tanks can be configured to have wastewater pass through an aerobic treatment train.

ATUs use biological processes to transform both dissolved and solid constituents into gases, cell mass, and non-degradable material. An important feature of the biological process is the synthesis and separation of microbial cells from the treated effluent.

The treatment process involves a variety of aerobic and facultative microorganisms living together that can decompose a broad range of materials. The organisms live in an aerobic environment where free oxygen is available for their respiration. ATUs can be used to remove substantial amounts of BOD5 and TSS that are not removed by simple sedimentation in a conventional septic system.

The biological process also involves the nitrification of ammonia in the wastewater and the reduction of pathogenic organisms. Nitrification is the breakdown of ammonia (NH3+) to nitrate (NO3-) by microorganisms in aerobic conditions.

ATUs, which are certified as Class I aerobic systems, treat wastewater well enough to be used in conjunction with spray systems, which distribute treated wastewater over lawns. Combined with disinfection, they are the most common way to treat wastewater for spray systems.

Process

The ATS process generally consists of the following phases:

- Pre-treatment stage to remove large solids and other undesirable substances.
- Aeration stage, where aerobic bacteria digest biological wastes.
- Settling stage allows undigested solids to settle. This forms a sludge that must be periodically removed from the system.
- Disinfecting stage, where chlorine or similar disinfectant is mixed with the water, to produce an antiseptic output. Another option is UV disinfection, where the water is exposed to UV light inside of a UV disinfection unit.

The disinfecting stage is optional, and is used where a sterile effluent is required, such as cases where the effluent is distributed above ground. The disinfectant typically used is tablets of calcium hypochlorite, which are specially made for waste treatment systems. The tablets are intended to break down quickly in sunlight. Stabilized forms of chlorine persist after the effluent is dispersed, and can kill plants in the leach field.

Since the ATS contains a living ecosystem of microbes to digest the waste products in the water, excessive amounts of items such as bleach or antibiotics can damage the ATS environment and reduce treatment effectiveness. Non-digestible items should also be avoided, as they will build up in the system and require more frequent sludge removal.

Types of Aerobic Treatment Systems

Small scale aerobic systems generally use one of two designs, fixed-film systems, or continuous flow, suspended growth aerobic systems (CFSGAS). The pre-treatment and effluent handling are similar for both types of systems, and the difference lies in the aeration stage.

Fixed Film Systems

Fixed film systems use a porous medium which provides a bed to support the biomass film that digests the waste material in the wastewater. Designs for fixed film systems vary widely, but fall into two basic categories (though some systems may combine both methods). The first is a system where the media is moved relative to the wastewater, alternately immersing the film and exposing it to air, while the second uses a stationary media, and varies the wastewater flow so the film is alternately submerged and exposed to air. In both cases, the biomass must be exposed to both wastewater and air for the aerobic digestion to occur. The film itself may be made of any suitable porous material, such as formed plastic or peat moss. Simple systems use stationary media, and rely on intermittent, gravity driven wastewater flow to provide periodic exposure to air and wastewater. A common moving media system is the rotating biological contactor (RBC), which uses disks rotating slowly on a horizontal shaft. Approximately 40 percent of the disks are submerged at any given time, and the shaft rotates at a rate of one or two revolutions per minute.

Continuous Flow, Suspended Growth Aerobic Systems

CFSGAS systems, as the name implies, are designed to handle continuous flow, and do not provide a bed for a bacterial film, relying rather on bacteria suspended in the wastewater. The suspension and aeration are typically provided by an air pump, which pumps air through the aeration chamber, providing a constant stirring of the wastewater in addition to the oxygenation. A medium to promote fixed film bacterial growth may be added to some systems designed to handle higher than normal levels of biomass in the wastewater.

Retrofit or Portable Aerobic Systems

Another increasingly common use of aerobic treatment is for the remediation of failing or failed anaerobic septic systems, by retrofitting an existing system with an aerobic feature. This class of product, known as aerobic remediation, is designed to remediate biologically failed and failing anaerobic distribution systems by significantly reducing the biochemical oxygen demand (BOD5) and total suspended solids (TSS) of the effluent. The reduction of the BOD5 and TSS reverses the developed bio-mat. Further, effluent with high dissolved oxygen and aerobic bacteria flow to the distribution component and digest the bio-mat.Doing so on single tank systems where solids do not have anywhere to settle, or there is no a clarifying area can do damage to the field lines as the solid matter is stirred up in the tank.

Composting Toilets

Composting toilets are designed to treat only toilet waste, rather than general residential waste water, and are typically used with water-free toilets rather than the flush toilets associated with the above types of aerobic treatment systems. These systems treat the waste as a moist solid, rather than in liquid suspension, and therefore separate urine from feces during treatment to maintain the correct moisture content in the system. An example of a composting toilet is the clivus multrum (Latin for 'inclined chamber'), which consists of an inclined chamber that separates urine and feces and a fan to provide positive ventilation and prevent odors from escaping through the toilet. Within the chamber, the urine and feces are independently broken down not only by aerobic bacteria, but also by fungi, arthropods, and earthworms. Treatment times are very long, with a minimum time between removals of solid waste of a year; during treatment the volume of the solid waste is decreased by 90 percent, with most being converted into water vapor and carbon dioxide. Pathogens are eliminated from the waste by the long durations in inhospitable conditions in the treatment chamber.

Comparison to Traditional Septic Systems

The aeration stage and the disinfecting stage are the primary differences from a traditional septic system; in fact, an aerobic treatment system can be used as a secondary treatment for septic tank effluent. These stages increase the initial cost of the aerobic system, and also the maintenance requirements over the passive septic system. Unlike many other biofilters, aerobic treatment systems require a constant supply of electricity to drive the air pump increasing overall system costs. The disinfectant tablets must be periodically replaced, as well as the electrical components (air compressor) and mechanical components (air diffusers). On the positive side, an aerobic system produces a higher quality effluent than a septic tank, and thus the leach field can be smaller than that of a conventional septic system, and the output can be discharged in areas too environmentally sensitive for septic system output. Some aerobic systems recycle the effluent through a sprinkler system, using it to water the lawn where regulations approve.

Effluent Quality

Since the effluent from an ATS is often discharged onto the surface of the leach field, the quality is very important. A typical ATS will, when operating correctly, produce an effluent with less than 30 mg/liter BOD5, 25 mg/L TSS, and 10,000 cfu/mL fecal coliform bacteria. This is clean enough that it cannot support a biomat or "slime" layer like a septic tank.

ATS effluent is relatively odorless; a properly operating system will produce effluent that smells musty, but not like sewage. Aerobic treatment is so effective at reducing odors, that it is the preferred method for reducing odor from manure produced by farms.

API Oil–water Separator

The API oil/water separator is a simple gravity separation device designed by using Stokes Law to define the rise velocity of oil droplets based on their density and size. The design of the separator is based on the specific gravity difference between the oil and the wastewater because that difference is usually smaller than the specific gravity difference between the suspended solids and water. Based on that design criterion, most of the suspended solids will settle to the bottom of the separator as a sediment layer, the oil will rise to top of the separator, and the wastewater will be the middle layer between the oil on top and the solids on the bottom.

API oil/water separators are designed to remove oil droplets down to 150 micron, which is not nearly enough removal necessary to stay compliant today. Most API oil/water separators are not going to do any better than 50 parts per million as opposed to most state & federal regulations that today require below 20 ppm.

Most industries have abandoned building these large gravity settling vessels because they are simply too costly to justify. Most are built of concrete and with the high cost of labor and the need to reinforce these tanks with steel, new construction costs rise quickly. Plus API units have an excessively large footprint and are far too much of a real estate burden to consider this as a viable option.

Those who have API oil/water separators acknowledge they are easy to operate and maintain on a day-to-day basis, but when it is time to clean out the tank, the cost to pump out the sludge and clean this large space is just a nightmare. Some units have a chain and scrapper mechanism. This helps move the solids to one side to be pumped out, but the automation piece for these are usually as much as the construction of the pit itself—and require significant maintenance as well.

Description of the Design and Operation

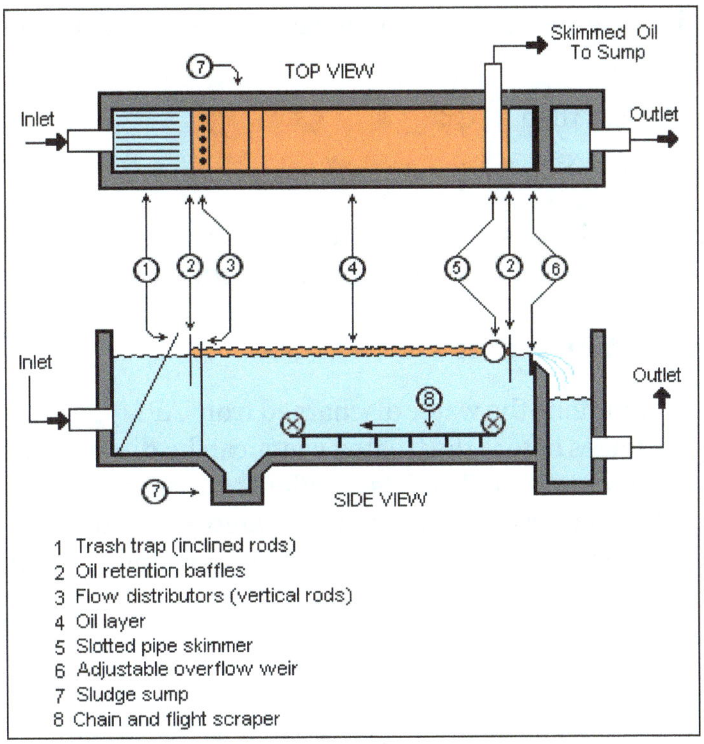

A typical gravimetric API separator.

The API separator is a gravity separation device designed using Stokes' law principles that define the rise velocity of oil droplets based on their density, size and water properties. The design of the separator is based on the specific gravity difference between the oil and the wastewater because that difference is much smaller than the specific gravity difference between the suspended solids and water. Based on that design criterion, most of the suspended solids will settle to the bottom of the separator as a sediment layer, the oil will rise to top of the separator, and the wastewater will be the middle layer between the oil on top and the solids on the bottom. The API Design Standards, when correctly applied, make adjustments to the geometry, design and size of the separator beyond simple Stokes Law principles. This includes allowances for water flow entrance and exit turbulence losses as well as other factors. API Specification 421 requires a minimum length to width ratio of 5:1 and minimum depth-to-width ratio of 0.3:0.5.

Typically in operation of API separators the oil layer, which may contain entrained water and attached suspended solids, is continually skimmed off. This removed oily layer may be re-processing to recover valuable products, or disposed of. The heavier bottom sediment layer is removed by a chain and flight scraper (or similar device) and a sludge pump.

Design Limitations

API design separators, and similar gravity tanks, are not intended to be effective when any of the following conditions apply to the feed conditions:

- Mean Oil droplets size in the feed is less than 150 micron.
- Oil density is greater than 925 kg/m3.
- Suspended solids are adhering to the oil meaning the 'effective' oil density is greater than 925 kg/m3.
- Water temperature less than 5 °C.
- There are high levels of dissolved hydrocarbons.

According to Stokes' Law, heavier oils require more retention time. In many cases where refineries have switched to heavier crude slates, the API separator's efficiency has declined.

Further Treatment of API Water Discharges

Because of performance limitations the water discharged from API type separators usually requires several further processing stages before the treated water can be discharged or reused. Further water treatment is designed to remove oil droplets smaller than 150 micron, dissolved materials and hydrocarbons, heavier oils or other contaminants not removed by the API. Secondary treatment technologies include dissolved air flotation (DAF), Anaerobic and Aerobic biological treatment, Parallel Plate Separators, Hydrocyclone, Walnut Shell Filters and Media filters.

Alternative Technologies

Plate separators, or Coalescing Plate Separators are similar to API separators, in that they are based on Stokes Law principles, but include inclined plate assemblies (also known as parallel packs). The underside of each parallel plate provides more surface for suspended oil droplets to coalesce into larger globules. Coalescing plate separators may not be effective in situation where water chemicals or suspended solids restrict or prevent oil droplets coalesce. In operation it is intended that sediment will slide down the topside of each parallel plate, however in many practical situations the sediment can adhere to the plates requiring periodic removal and cleaning. Such separators still depend upon the specific gravity between the suspended oil and the water. However, the parallel plates can enhance the degree of oil-water separation for oil droplets above 50 micron in size. Alternatively parallel plate separators are added to the design of API Separators and require less space than a conventional API separator to achieve a similar degree of separation.

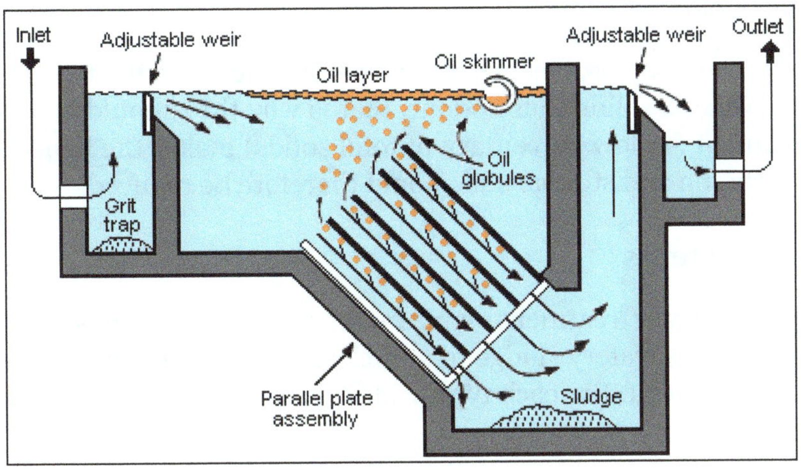

A typical parallel plate separator.

Other Oil–water Separation Applications

There are other applications requiring oil-water separation. For example:

- Oily water separators (OWS) for separating oil from the bilge water accumulated in ships as required by the international MARPOL Convention.
- Oil and water separators are commonly used in electrical substations. The transformers found in substations use a large amount of oil for cooling purposes. Moats are constructed surrounding unenclosed substations to catch any leaked oil, but these will also catch rainwater. Oil and water separators therefore provide a quicker and easier cleanup of an oil leak.

Sewage Sludge Treatment

Sewage sludge consists of water and solids that can be divided into mineral and organic solids. The quantity and characteristics of sludge depend very much on the treatment processes. Most of the pollutants that enter the wastewater get adsorbed to the sewage sludge. Therefore, sewage sludge contains pathogens (and heavy metals, many organic pollutants pesticides, hydrocarbons etc. if the sewage contains industrial influence). Sludge is, however, rich in nutrients such as nitrogen and phosphorous and contains valuable organic matter that is useful if soils are depleted or subject to erosion.

Options for sludge treatment include stabilisation, thickening, dewatering, drying. Sewage sludge is stabilised to reduce pathogens, to eliminate offensive odours and to inhibit, reduce or eliminate the potential for putrefaction. Moreover, stabilisation is used for volume reduction, production of usable gas (methane), and improving the dewaterability of sludge. Thickening, dewatering, drying are used to remove water from sewage sludge. Several techniques are used in dewatering devices for removing moisture. A technique close to nature and very effective is dewatering in drying beds. The principal advantages of drying beds are low costs, infrequent attention required, and high solids content in the dried product, especially in arid climates. Disadvantages are the large space required, effects of climatic changes on drying characteristics, labour-intensive sludge removal, insects and potential odours.

The purpose of sludge treatment is, besides hygienisation, to change the figures in the table above. Easily biodegradable volatile solids cause odour, that is why they should be reduced by stabilisation. A high water content (→ low percentage of total solids) makes the handling difficult, causes high cost for transportation and storage and should therefore be reduced.

Sludge Pumping Systems

Sludge produced in wastewater treatment plants must be conveyed from point to point in the plant in conditions ranging from a watery sludge to a thick sludge. Sludge may also be pumped off-site for long distances for treatment. For each type of sludge and pumping application a different type of pump may be needed.

Pumps can generally be divided into centrifugal pumps (with different impellers) and displacement pumps (progressive cavity, rotary lobe, piston). Centrifugal pumps are suitable for high flow rates and low solid contents, the problem is choosing a proper size. At any given speed, centrifugal pumps operate well only if the pumping head is within a relatively narrow range. The variable nature of sludge, however, causes pumping heads to change. The selected pump must have sufficient clearance to pass the solids without clogging. Usual centrifugal pumps can cause the break up of flocculent particles in activated sludge. Good experiences in this respect have been made with screw-shaped impellers.

Table: Suitable pumps for sludge

Type of pump	Applicable for:	Advantages	Disadvantages
Centrifugal pumps	thin sludge (max. 2.5 - 3%)		
Nonclog	activated sludge	high volume, good efficiency	potential clogging (rags etc.)
Recessed impeller	sludge with solids or grit		lower efficiency
Chopper	primary sludge	reduces clogging	lower efficiency
Progressing cavity pump	thickened sludge dewatered sludge	defined flow rates acts as check valve	can run dry grit can cause high stator wear
Rotary lobe pump	thickened sludge dewatered sludge	defined flow rates acts as check valve	grit can cause high lobe wear
Piston pump	thickened sludge dewatered sludge	high pressure	discontinuous flow

Screw pump	activated sludge	good efficiency	limited height high capital costs space requirement
Screw	dried sludge		
conveyor belt	dewatered sludge dried sludge	good efficiency	high capital costs space requirement

Displacement pumps can convey dewatered sludge up to 30% TS. Displacement pumps have a fixed ratio between revolutions and flow rate. Any pipeline obstruction causes damages to the pipeline or pump. Generally check valves are not necessary. For primary sludge, a grinder normally proceeds progressive cavity pumps.

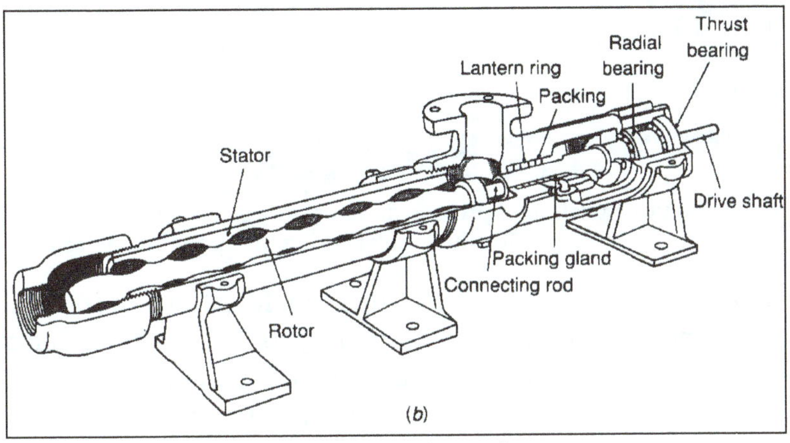

Scheme of a progressive cavity pump.

Treatment Processes

Thickening

A sewage sludge thickener.

Thickening is often the first step in a sludge treatment process. Sludge from primary or secondary clarifiers may be stirred (often after addition of clarifying agents) to form larger, more rapidly settling aggregates. Primary sludge may be thickened to about 8 or 10 percent solids, while secondary sludge may be thickened to about 4 percent solids. Thickeners often resemble a clarifier with the addition of a stirring mechanism. Thickened sludge with less than ten percent solids may receive additional sludge treatment while liquid thickener overflow is returned to the sewage treatment process.

Dewatering

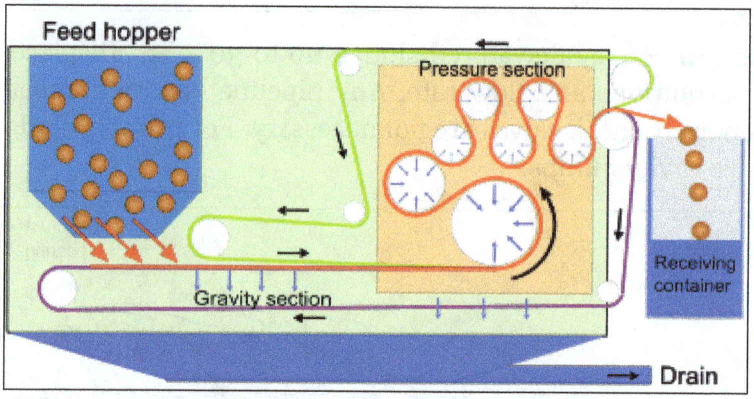

Schematic of a belt filter press to dewater sewage sludge. Filtrate is extracted initially by gravity, then by squeezing the cloth through rollers.

Water content of sludge may be reduced by centrifugation, filtration, and/or evaporation to reduce transportation costs of disposal, or to improve suitability for composting. Centrifugation may be a preliminary step to reduce sludge volume for subsequent filtration or evaporation. Filtration may occur through underdrains in a sand drying bed or as a separate mechanical process in a belt filter press. Filtrate and centrate are typically returned to the sewage treatment process. After dewatering sludge may be handled as a solid containing 50 to 75 percent water. Dewatered sludges with higher moisture content are usually handled as liquids.

Sidestream Treatment Technologies

Sludge treatment technologies that are used for thickening or dewatering of sludge have two products: the thickened or dewatered sludge, and a liquid fraction which is called sludge treatment liquids, sludge dewatering streams, liquors, centrate (if it stems from a centrifuge), filtrate (if it stems from a belt filter press) or similar. This liquid requires further treatment as it is high in nitrogen and phosphorus, particularly if the sludge has been anaerobically digested. The treatment can take place in the sewage treatment plant itself (by recycling the liquid to the start of the treatment process) or as a separate process.

Phosphorus Recovery

One method for treating sludge dewatering streams is by using a process that is also used for phosphorus recovery. Another benefit for sewage treatment plant operators of treating sludge dewatering streams for phosphorus recovery is that it reduces the formation of obstructive struvite scale in pipes, pumps and valves. Such obstructions can be a maintenance headache particularly for biological nutrient removal plants where the phosphorus content in the sewage sludge is elevated. For example, the Canadian company Ostara Nutrient Recovery Technologies is marketing a process based on controlled chemical precipitation of phosphorus in a fluidized bed reactor that recovers struvite in the form of crystalline pellets from sludge dewatering streams. The resulting crystalline product is sold to the agriculture, turf and ornamental plants sectors as fertiliser under the registered trade name "Crystal Green".

Digestion

Many sludges are treated using a variety of digestion techniques, the purpose of which is to reduce the amount of organic matter and the number of disease-causing microorganisms present in the solids. The most common treatment options include anaerobic digestion, aerobic digestion, and composting. Sludge digestion offers significant cost advantages by reducing sludge quantity by nearly 50% and providing biogas as a valuable energy source.

Anaerobic Digestion

Anaerobic digestion is a bacterial process that is carried out in the absence of oxygen. The process can either be *thermophilic* digestion, in which sludge is fermented in tanks at a temperature of 55 °C, or *mesophilic*, at a temperature of around 36 °C. Though allowing shorter retention time (and thus smaller tanks), thermophilic digestion is more expensive in terms of energy consumption for heating the sludge.

Mesophilic anaerobic digestion (MAD) is also a common method for treating sludge produced at sewage treatment plants. The sludge is fed into large tanks and held for a minimum of 12 days to allow the digestion process to perform the four stages necessary to digest the sludge. These are hydrolysis, acidogenesis, acetogenesis, and methanogenesis. In this process the complex proteins and sugars are broken down to form more simple compounds such as water, carbon dioxide, and methane.

Anaerobic digestion generates biogas with a high proportion of methane that may be used to both heat the tank and run engines or microturbines for other on-site processes. Methane generation is a key advantage of the anaerobic process. Its key disadvantage is the long time required for the process (up to 30 days) and the high capital cost. Many larger sites utilize the biogas for combined heat and power, using the cooling water from the generators to maintain the temperature of the digestion plant at the required 35 ± 3 °C. Sufficient energy can be generated in this way to produce more electricity than the machines require.

Aerobic Digestion

Aerobic digestion is a bacterial process occurring in the presence of oxygen resembling a continuation of the activated sludge process. Under aerobic conditions, bacteria rapidly consume organic matter and convert it into carbon dioxide. Once there is a lack of organic matter, bacteria die and are used as food by other bacteria. This stage of the process is known as *endogenous respiration*. Solids reduction occurs in this phase. Because the aerobic digestion occurs much faster than anaerobic digestion, the capital costs of aerobic digestion are lower. However, the operating costs are characteristically much greater for aerobic digestion because of energy used by the blowers, pumps and motors needed to add oxygen to the process. However, recent technological advances include non-electric aerated filter systems that use natural air currents for the aeration instead of electrically operated machinery.

Aerobic digestion can also be achieved by using diffuser systems or jet aerators to oxidize the sludge. Fine bubble diffusers are typically the more cost-efficient diffusion method, however, plugging is typically a problem due to sediment settling into the smaller air holes. Coarse bubble diffusers are more commonly used in activated sludge tanks or in the flocculation stages. A key component for selecting diffuser type is to ensure it will produce the required oxygen transfer rate.

Composting

Composting is an aerobic process of mixing sewage sludge with agricultural byproduct sources of carbon such as sawdust, straw or wood chips. In the presence of oxygen, bacteria digesting both the sewage sludge and the plant material generate heat to kill disease-causing microorganisms and parasites.[:20] Maintenance of aerobic conditions with 10 to 15 percent oxygen requires bulking agents allowing air to circulate through the fine sludge solids. Stiff materials like corn cobs, nut shells, shredded tree-pruning waste, or bark from lumber or paper mills better separate sludge for ventilation than softer leaves and lawn clippings. Light, biologically inert bulking agents like shredded tires may be used to provide structure where small, soft plant materials are the major source of carbon.

Uniform distribution of pathogen-killing temperatures may be aided by placing an insulating blanket of previously composted sludge over aerated composting piles. Initial moisture content of the composting mixture should be about 50 percent; but temperatures may be inadequate for pathogen reduction where wet sludge or precipitation raises compost moisture content above 60 percent. Composting mixtures may be piled on concrete pads with built-in air ducts to be covered by a layer of unmixed bulking agents. Odors may be minimized by using an aerating blower drawing vacuum through the composting pile via the underlying ducts and exhausting through a filtering pile of previously composted sludge to be replaced when moisture content reaches 70 percent. Liquid accumulating in the underdrain ducting may be returned to the sewage treatment plant; and composting pads may be roofed to provide better moisture content control.

After a composting interval sufficient for pathogen reduction, composted piles may be screened to recover undigested bulking agents for re-use; and composted solids passing through the screen may be used as a soil amendment material with similar benefits to peat. The optimum initial carbon-to-nitrogen ratio of a composting mixture is between 26-30:1; but the composting ratio of agricultural byproducts may be determined by the amount required to dilute concentrations of toxic chemicals in the sludge to acceptable levels for the intended compost use. Although toxicity is low in most agricultural byproducts, suburban grass clippings may have residual herbicide levels detrimental to some agricultural uses; and freshly composted wood byproducts may contain phytotoxins inhibiting germination of seedlings until detoxified by soil fungi.

Incineration

Incineration of sludge is less common because of air emissions concerns and the supplemental fuel (typically natural gas or fuel oil) required to burn the low calorific value sludge and vaporize residual water. On a dry solids basis, the fuel value of sludge varies from about 9,500 British thermal units per pound (980 cal/g) of undigested sewage sludge to 2,500 British thermal units per pound (260 cal/g) of digested primary sludge. Stepped multiple hearth incinerators with high residence time and fluidized bed incinerators are the most common systems used to combust wastewater sludge. Co-firing in municipal waste-to-energy plants is occasionally done, this option being less expensive assuming the facilities already exist for solid waste and there is no need for auxiliary fuel. Incineration tends to maximize heavy metal concentrations in the remaining solid ash requiring disposal; but the option of returning wet scrubber effluent to the sewage treatment process may reduce air emissions by increasing concentrations of dissolved salts in sewage treatment plant effluent.

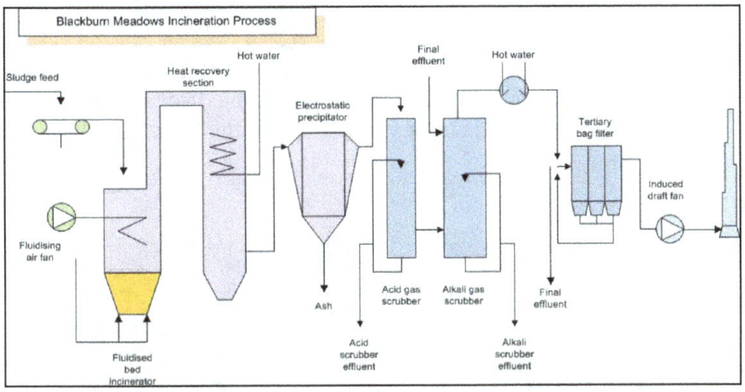

Sludge incineration process schematic (note the emphasis on air quality control).

Drying Beds

Simple sludge drying beds are used in many countries, particularly in developing countries, as they are a cheap and simple method to dry sewage sludge. Drainage water must be captured; drying beds are sometimes covered but usually left uncovered. Mechanical devices to turn over the sludge in the initial stages of the drying process are also available on the market.

Drying beds are typically composed of four layers consisting of gravel and sand. The first layer is coarse gravel that is 15 to 20 centimeters thick. Followed by fine gravel that is 10 centimeters thick. The third layer is sand that can be between 10 to 15 centimeters and serves as the filter between the sludge and gravel. Sludge dries up and water percolates to the first layer that is collected at the drainage pipe that is beneath all layers.

Sewage sludge after drying in a sludge drying bed.

This simple evaporative sludge drying bed near Damascus in Syria illustrates the initial consistency of primary sludge being discharged from the primary settling tank via the pipe in the foreground.

Emerging Technologies

The thermal hydrolysis system at the Blue Plains treatment plant in Washington, D.C. is the largest in the world.

- Phosphorus recovery from sewage sludge or from sludge dewatering streams is receiving increased attention particularly in Sweden, Germany and Canada, as phosphorus is a limited resource (a concept also known as "peak phosphorus") and is needed as fertilizer to feed a growing world population. Phosphorus recovery methods from wastewater or sludge can be categorized by the origin of the used matter (wastewater, sludge liquor, digested or non-digested sludge, ash) or by the type of recovery processes (precipitation, wet-chemical extraction and precipitation, thermal treatment). Research on phosphorus recovery methods from sewage sludge has been carried out in Sweden and Germany since around 2003, but the technologies currently under development are not yet cost effective, given the current price of phosphorus on the world market.
- The Omni Processor is a process that is currently under development that treats sewage sludge and can generate a surplus of electrical energy if the input materials have the right level of dryness.
- Thermal depolymerization produces light hydrocarbons from sludge heated to 250 °C and compressed to 40 MPa.

- Thermal hydrolysis is a two-stage process combining high-pressure boiling of sludge, followed by a rapid decompression. This combined action sterilizes the sludge and makes it more biodegradable, which improves digestion performance. Sterilization destroys pathogens in the sludge resulting in it exceeding the stringent requirements for land application (agriculture). Thermal hydrolysis systems are operating at sewage treatment plants in Europe, China and North America, and can generate electricity as well as high quality sludge.

Disposal or use as Fertilizer

When a liquid sludge is produced, further treatment may be required to make it suitable for final disposal. Sludges are typically thickened and/or dewatered to reduce the volumes transported off-site for disposal. Processes for reducing water content include lagooning in drying beds to produce a cake that can be applied to land or incinerated; pressing, where sludge is mechanically filtered, often through cloth screens to produce a firm cake; and centrifugation where the sludge is thickened by centrifugally separating the solid and liquid. Sludges can be disposed of by liquid injection to land or by disposal in a landfill.

There is no process which completely eliminates the need to dispose of treated sewage sludge.

Much sludge originating from commercial or industrial areas is contaminated with toxic materials that are released into the sewers from the industrial processes. Elevated concentrations of such materials may make the sludge unsuitable for agricultural use and it may then have to be incinerated or disposed of to landfill.

Examples

Edmonton, Alberta, Canada

The Edmonton Composting Facility, in Edmonton, Alberta, Canada, is the largest sewage sludge composting site in North America.

New York City, U.S.

Sewage sludge can be superheated and converted it into pelletized granules that are high in nitrogen and other organic materials. In New York City, for example, several sewage treatment plants have dewatering facilities that use large centrifuges along with the addition of chemicals such as polymer to further remove liquid from the sludge. The product which is left is called "cake," and that is picked up by companies which turn it into fertilizer pellets. This product, also called biosolid, is then sold to local farmers and turf farms as a soil amendment or fertilizer, reducing the amount of space required to dispose of sludge in landfills.

Southern California, U.S.

In the very large metropolitan areas of southern California inland communities return sewage sludge to the sewer system of communities at lower elevations to be reprocessed at a few very large treatment plants on the Pacific coast. This reduces the required size of interceptor sewers and allows local recycling of treated wastewater while retaining the economy of a single sludge processing facility and is an example of how sewage sludge can help solve an energy crisis.

Controversy

Some campaigners perceive sewage sludge treatment as a problem and a danger to the environment - largely because systems in most industrialised countries mix industrial wastes with household sewerage. This has led some to claim that the term "biosolids" was created by the sewage treatment industry in order to take the focus off the origins of the material to make reuse more acceptable to the public, and some studies have suggested that this is in fact a form of propaganda.

Coagulation Water Treatment

The coagulation process involves adding iron or aluminum salts, such as aluminum sulphate, ferric sulphate, ferric chloride or polymers, to the water. These chemicals are called coagulants, and have a positive charge. The positive charge of the coagulant neutralizes the negative charge of dissolved and suspended particles in the water. When this reaction occurs, the particles bind together, or coagulate (this process is sometimes also called flocculation). The larger particles, or floc, are heavy and quickly settle to the bottom of the water supply. This settling process is called sedimentation. The following diagram illustrates the basic reactions and processes that occur during coagulation.

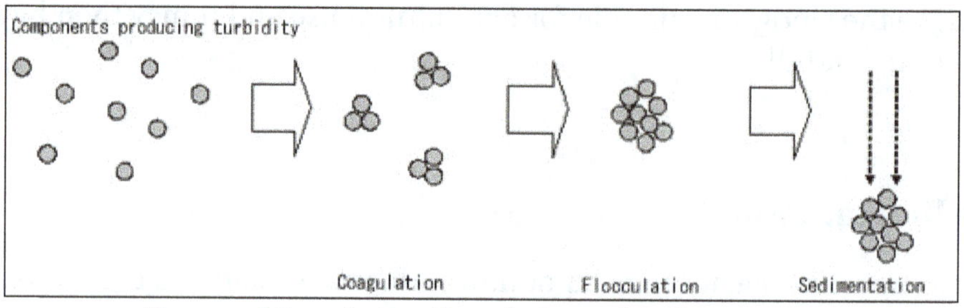

Process of Coagulation, Flocculation and Sedimentation.

The chart below shows the length of time that is required for particles of different sizes to settle through the water.

Diameter of Particle	Type of Particle	Settling time through 1 m. of water
10mm	Gravel	1 seconds
1mm	Sand	10 seconds
0.1mm	Fine Sand	2 minutes
10 micron	Protozoa, Algae, Clay	2 hours
1 micron	Bacteria, Algae	8 days
0.1 micron	Viruses, Colloids	2 years
10 nm	Viruses, Colloids	20 years
1 nm	Viruses, Colloids	200 years

Settling Time for Particles of Various Diameters.

In a water treatment facility, the coagulant is added to the water and it is rapidly mixed, so that the coagulant is circulated throughout the water. The coagulated water can either be filtered directly through a medium filter (such as sand and gravel), a microfiltration or ultrafiltration membrane, or it can be moved to a settling tank. In a settling tank, or clarifier, the heavy particles settle to the bottom and are removed, and the water moves on to the filtration step of the treatment process.

Coagulation can successfully remove a large amount of organic compounds, including some dissolved organic material, which is referred to as Natural Organic Matter (NOM) or Dissolved Organic Carbon (DOC). Coagulation can also remove suspended particles, including inorganic precipitates, such as iron. A large amount of DOC can give water an unpleasant taste and odour, as well as a brown discolouration. While coagulation can remove particles and some dissolved matter, the water may still contain pathogens. In an international report published in 1998, it was found that coagulation and sedimentation can only remove between 27 and 84 percent of viruses and between 32 and 87 percent of bacteria. Usually, the pathogens that are removed from the water are removed because they are attached to the dissolved substances that are removed by coagulation. In the picture below, the coagulants have been added to the water, and the particles are starting to bind together and settle to the bottom.

Coagulation jar test in a water treatment plant.

As coagulation does not remove all of the viruses and bacteria in the water, it cannot produce safe drinking water. It is, however, an important primary step in the water treatment process, because coagulation removes many of the particles, such as dissolved organic carbon, that make water difficult to disinfect. Because coagulation removes some of the dissolved substances, less chlorine must be added to disinfect the water. A municipal water treatment plant can save money by using less chlorine, and the water will be safer, because trihalomethanes (THMs) are a dangerous by-product that results from the reaction of chlorine with NOM.

With accurate dosages and proper application, the residuals of the added chemicals generally do not pose a problem. Residuals are the by-products that remain in the water after substances are added and reactions occur within the water. The particular residuals depend on the coagulant that is used. If ferric sulphate is used, iron and sulphate are added to the water. If ferric chloride is used, iron and chloride are added. And if aluminum sulphate is used, aluminum and sulphate are added. The majority of municipal water treatment plants use aluminum sulphate as the coagulation chemical. Generally, water treatment facilities have the coagulation process set up so that the coagulant chemicals are removed with the floc. However, it is widely accepted that treatment facilities that use aluminum based coagulants often have higher levels of aluminum in their treated water, but not by much. In Canada, Health Canada has different guideline parameters for treatment facilities that do not use aluminum based coagulants (i.e., 0.1 mg/L) and those that do (i.e., 0.2 mg/L).

Factors

Coagulation is affected by the type of coagulant used, its dose and mass; pH and initial turbidity of the water that is being treated; and properties of the pollutants present. The effectiveness of the coagulation process is also affected by pretreatments like oxidation.

Mechanism

In a colloidal suspension, particles will settle very slowly or not at all because the colloidal particles carry surface electrical charges that mutually repel each other. A coagulant (typically a metallic salt) with the opposite charge is added to the water to overcome the repulsive charge and "destabilize" the suspension. For example, the colloidal particles are negatively charged and alum is added as a coagulant to create positively charged ions. Once the repulsive charges have been neutralized (since opposite charges attract), the van der Waals force will cause the particles to cling together (agglomerate) and form micro floc.

Determining Coagulant Dose

Jar Test

Jar test for coagulation.

The dose of the coagulant to be used can be determined via the Jar Test. The jar test involves exposing same volume samples of the water to be treated to different doses of the coagulant and then simultaneously mixing the samples at a constant rapid mixing time. The microfloc formed after coagulation further undergoes flocculation and is allowed to settle. Then the turbidity of the samples are measured and the dose with the lowest turbidity can be said to be optimum.

Streaming Current Detector

An automated device for determining the coagulant dose is the Streaming Current Detector (SCD).

The SCD measures the net surface charge of the particles and shows a streaming current value of 0 when the charges are neutralized (cationic coagulants neutralize the anionic colloids). At this value (0), the coagulant dose can be said to be optimum.

Limitations

Coagulation itself results in the formation of floc but flocculation is required to help the floc further aggregate and settle. The coagulation-flocculation process itself removes only about 60%-70% of Natural Organic Matter (NOM) and thus, other processes like oxidation, filtration and sedimentation are necessary for complete raw water or wastewater treatment. Coagulant aids (polymers that bridge the colloids together) are also often used to increase the efficiency of the process.

Factors Affecting Coagulation Water Treatment

The process of coagulation of water depends on various factors like pH of the medium, temperature of water, coagulant feed concentration, coagulant dosage, type of coagulant, mass and initial turbidity. Moreover it is also depends on pre-treatment and type of pollutants present.

Effect of pH on Coagulation

pH affects on the activities of coagulants. The optimum pH for alum coagulation is 6 to 7.5 whereas 5.0 to 8.0 are for iron. If the alkalinity is lower or higher, then the floc does not form properly. As a result, more coagulant is consumed. In this case, it is beneficial to correct the pH by adding acid or base.

Temperature

Temperature is another factor for coagulation water treatment process. It is more significant at lower turbidity. In case of alum, at low temperatures aluminium hydroxide form a strongly hydrated and very stable sol. So in winter season high coagulant are consumed. When the temperature becomes below the 5^0C, then alum or ferric salts do not work properly. So it should be consider another coagulant like polyaluminum chloride (PACl).

Type of Pollutants

The salt composition of soft water and hard water are not same. Hard water contains Ca^{2+} and Mg^{2+} ions. They can alter the charge on the colloidal particles.

Optimum Dosage

It is very significant to determine the optimum dosage of a coagulant which will give the maximum clarifying effect. Insufficient amount of coagulant cannot able to destabilize properly of the colloidal particles. On the other hand higher dosage can cause excessive sludge production, corrosion and loss of money.

Type of Coagulant

All the coagulants are not suitable for all cases. Different temperatures, pH, type of medium may vary the effectiveness of the coagulant. At lower temperature the polyaluminum chloride (PACl) may be more effective than the traditional coagulants like alum or iron salt. Same way, some pH range can be beneficial to use iron salt instead of alum.

Coagulation Jar Test

You can determine optimum process condition like dosage, pH by jar test experiments. Generally, it consist several jars filled with equal volume of water. Then test for various dosage of coagulant, pH etc.

Rotating Biological Contactor

Rotating biological contactors (RBC) are a conventional aerobic biological wastewater treatment unit. Conventional biological treatment means activated sludge systems and fixed film systems such as trickling filters, or RBC (NOLDE 1996). The advantage of all these systems is that they are compact (i.e. in densely populated urban settings) and that they efficiently reduce organic matter (JENSSEN 2004). However, they are high-tech and generally require skilled staff for construction as well as for operation. RBC can treat domestic black- or greywater and any other low- or high-strength biodegradable wastewater (e.g. industrial wastewater from food processors or paper mills). They have been found to be particularly effective for decentralised applications (on the level of a small to medium community or industry/institution), where electricity and skilled staff are available.

Treatment Process and Basic Design Principles

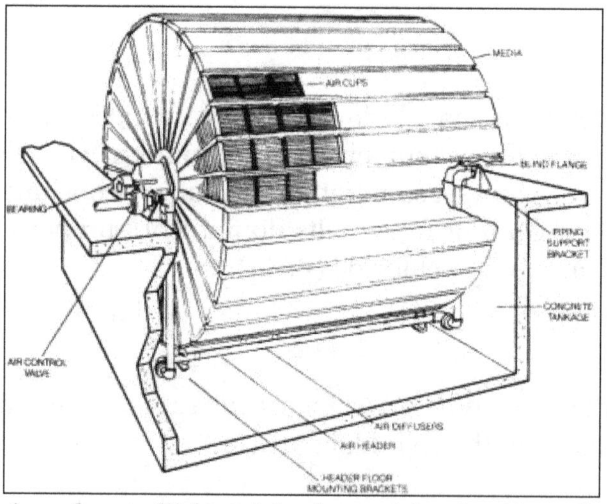

The disc is made out of light-weighted material such and usually ridged, corrugated, or lattice-like to make as much as surface available for the biofilm to attach.

A series of circular lightweight rotating discs are mounted on a shaft through which wastewater flows. The partially submerged discs rotate through the wastewater slowly. The disks are most commonly made of high-density plastic sheets (e.g. Polyethylene, polystyrene or polyvinylchloride) and are usually ridged, corrugated, or lattice-like to increase the specific surface area (NSFC 2004). The surface of the disks provides an attachment site for bacteria and as the discs rotate, a film of biomass grows on their surfaces (NSFC 2004; WSP 2008). This biofilm is alternately exposed to either the air or the wastewater as it rotates. The oxygen necessary for the growth of these microorganisms is obtained by adsorption from the air as the biofilm on the disk is rotated out of the liquid (CRITES & TCHOBANOGLOUS 1998; SANIMAS 2005). As the biofilm passes through the liquid phase, nutrients and organic pollutants are taken up. All oxygen, nutrients and organic pollutants are necessary for the growth of the microorganism and the conversion of the organic matter to CO_2. Nitrogen is removed by nitrification and subsequent denitrification transforming it to gaseous N_2, which is released to the air. The process is optimised by adjusting the speed of rotation and the depth of submergence (METCALF & EDDY Inc. 2003). In some designs, air is added to the bottom of the tank to provide additional oxygen in case of high-strength influents (CRITES & TCHOBANOGLOUS 1998).

The submerging level varies from 40 to 80 % (CRITES & TCHOBANOGLOUS 1998) and a usual rotating speed is 1 to 2 rpm (U.S. EPA 1980). The common disc diameter is between 0.6 and 3 m (SANIMAS 2005). The degradation process is similar to the one in a trickling filter with a high rate of recirculation (CRITES & TCHOBANOGLOUS 1998). The higher contact time in RBCs due to rotation allows up to 8 to 10 times higher levels of treatment than in trickling filters (WSP 2008). Also because the rotation allows both optimum wetting and oxygen supply, RBCs are generally more reliable than other fixed-film processes. Additionally, the disc design is made in such a way that large amounts of biofilm can attach, which means that there is a large amount of biological mass present to degrade the pollutants (WSP 2007). The large amount of biomass and the stability of contact also results in an improved stability and a reduced susceptibility to changes in hydraulic or organic loading compared to conventional activated sludge processes (WSP 2007). As for all fixed-film processes, primary settling and/or screening is required for the removal of grit, debris, and excessive oil (U.S. EPA 1980, WSP 2008). Such primary treatments are typically septic tanks, Imhoff tanks or anaerobic reactors. To remove sloughing sludge, a post-settling unit (i.e. a clarifier) is also required.

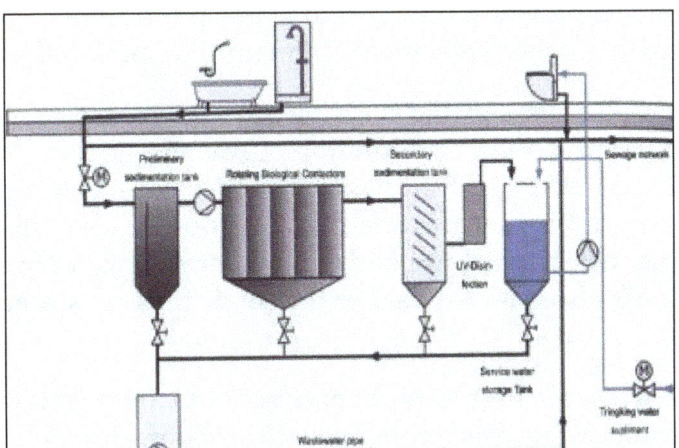

RBCs are a secondary treatment and as for all fixed-film processes, primary settling as well as sedimentation of sloughed sludge in a tertiary clarifier is required (example: greywater treatment in Germany).

The performance of RBC systems depends on the design, the temperature, the concentration of the pollutants, the rotating velocity and the hydraulic retention time. RBCs can achieve biological oxygen demand (BOD) reductions of 80 to 90 % (SANIMAS 2005; WSP 2007; WSP 2008). The removal of nitrogen (which is mostly present as ammonia) by nitrification and subsequent denitrification is also high, because both aerobic nitrifying bacteria and anaerobic denitrifying bacteria can simultaneously live in the attached biofilm (HOCHHEIMER 1998), depending on weather they are situated on the bottom of the film, close to the disc support (and thus in anaerobic or anoxic conditions) or at the top of the film exposed to the air.

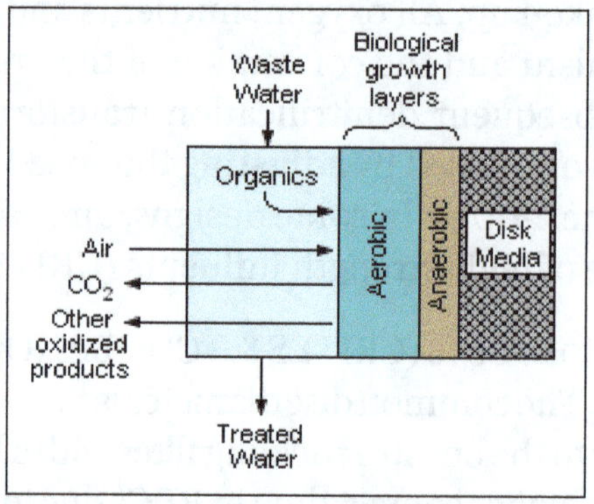

Both aerobic and anaerobic microorganisms can live in the biofilm and contribute to the removal of pollutant form the water.

Some other microorganisms which can transform ammonia (NH3) in one single step to gaseous N2 under anaerobic conditions have also been discovered in biofilms growing on RBC. These bacteria were called annamox and resulted in the development of innovative aerobic ammonia removal and wastewater treatment processes.

Little is known about the removal of phosphorus in RBCs, but it can be presumed that large parts of the phosphorus present is either accumulated in the biofilm or in the settled and collected sludge.

RBCs can be arranged in a variety of ways depending on specific effluent characteristics and the secondary clarifier design (e.g. specifically for BOD removal or nitrification, NSFC 2004).

Excess biomass sloughs off the discs by the shearing forces exerted as the discs rotate, combined with the force of gravity (WSP 2008). The rotation movement helps to keep sloughed solids in suspension so they can be carried to a clarifier (gravity settler) for secondary settling. The collected sludge in the clarifier requires further treatment (WSP 2008) for stabilisation, such as anaerobic digestion, composting, constructed wetlands, ponds or drying. Very often in small installations, accumulated sludge is also directed back to the septic tank for storage and partial digestion (U.S.EPA 1980).

Effluents from RBC do not contain high levels of nutrients and are therefore not particularly interesting for agriculture, although they constitute a source of water. However, due to reduced removal of microorganisms (1 to 2 log units, U.S. EPA 2002), RBC effluents require a further treatment, such as sand filtration, constructed wetlands or another form of disinfection (e.g. chemical disinfection or UV disinfection).

RBCs are usually designed on the basis of hydraulic and organic loadings derived from pilot plants and other full-scale installation (WSP 2007). Hydraulic retention times (HRTs) generally lye within some hours up to two days.

Even though RBCs are resistant to shock loading, long-term high organic loading may cause anaerobic conditions, resulting in odour and poor treatment performance (U.S.EPA 1980).

Example of an underground RBC for the decentralised treatment of domestic blackwater.

Recirculation is not normally practised in package fixed-film systems since it adds to the degree of complexity and is energy and maintenance intensive. However, recirculation may be desirable in certain applications where minimum wetting rates are required for optimal performance (U.S.EPA 1980).

Units may be installed at or below ground depending upon site topography and other adjacent treatment processes. Access to all moving parts and controls is required, and proper venting of the units is paramount, especially if natural ventilation is being used to supply oxygen (U.S.EPA 1980).

RBCs are often covered with a fibreglass housing to protect the disks from sunlight, wind, rain and low temperatures as performance of RCS drops considerably at air temperatures below 12°C (U.S.EPA 2002; NSFC 2004).

Costs Considerations

Observed costs for RBCs are highly variable depending on climate and location. Generally, RBCs involve high capital costs as not all materials may be locally available and motor and special material for rotation is required. Another cost factor may be manufacture and implementation, which requires skilled experts (SANIMAS 2005).

Operation and maintenance costs are also relatively high, because operation requires a continuous electricity supply and supervision requires semi-skilled labour (U.S.EPA 1980) and professional operator (SANIMAS 2005).

Operation and Maintenance

Large-scale RBCs are often covered to protect them form cold temperatures, rain, wind and sun. Sometimes, artificial aeration is required to keep the process aerobic when the systems are covered.

During operation, the system must be supervised by professional operators (SANIMAS 2005). Maintenance includes lubrication of moving parts, motors and bearings; replacing seals, motors, servicing bearings; and cleaning the attached-growth media (spray-washing of discs and purging of settled sludge) (METCALF & EDDY Inc. 2003; WSP 2007). The discs may be also checked for debris accumulation, ponding and excessive or not sufficient biomass accumulation (U.S.EPA 2002).

Although fixed film units such as RBC and trickling filters are operation- and maintenance-intensive, they do not require seeding with bacterial cultures (as do anaerobic processes such as anaerobic baffled reactors, septic tanks, upflow anaerobic sludge blanket reactors or anaerobic digesters) and the start-up phase is therefore considerably shorter. However, it takes 6 to 12 weeks for the biofilm to establish for a good treatment performance (U.S.EPA 2002).

Health Aspects

For correct operation, RBCs are covered and thus protected from contact with humans or animals. However, excess sludge as well as the effluent require post-treatment to remove pathogenic microorganisms.

In any case (i.e. for operation and maintenance) direct contact with the biomass growing on the discs, the effluent or the sludge should be avoided.

For discharge or reuse, a treatment unit allowing further pathogen removal should be considered.

Applicability

RBCs can achieve a high removal of biodegradable organic pollutants form domestic black- or greywater as well as from high-strength industrial wastewater (e.g. from dairies, bakeries, food processors, pulp, paper mills, WSP 2008). A great variety of applications are known, either as post-treatment for activated sludge processes in conventional domestic wastewater treatment plants, or for decentralized application at the level of small to medium-sized communities, industries or institutions (WSP 2007; WSP 2008).

They are adapted for urban areas mostly: land requirements are low, but continuous and consequent energy supply as well as semi-skilled labour are indispensable.

Some of the material may be locally available, however, the system can only be planned and implemented by experts (SANIMAS 2005).

Bar Screen

Bar screen and screening water treatment is the first process unit operation used at wastewater treatment plants. Screening removes objects such as rags, paper, plastics and metals to prevent damage and clogging of downstream equipment and piping.Cleaning frequency depends on the characteristics of the wastewater entering a plant.

Bar Screen Operation and Maintenance Considerations

- Check and clean the bar screen at frequent intervals.
- Do not allow solids to overflow /escape from the screen bar screen.
- Ensure no large gaps are formed due to the breakage of the screening water treatment.
- Replace breakage bar screen immediately.
- Mechanically cleaned screening system to remove larger materials because they reduce labor cost and they improve flow conditions and screening capture.
- Mechanically cleaned bar screen should have a standby screen to put in operation when the primary screening device is out of service.

Trouble Shooting of Screening Water Treatment

S.NO	PROBLEM	CAUSE
1	Large particles pass through and check the pumps	Poor operation /screen damaged
2	Up stream water levels is much higher than down stream level	Poor operation (inadequate cleaning)
3	Excessive collection of trash on screening water treatment	Poor operation
4	Excessive odor	Poor operation /trash disposal practices.

Daily and Weekly Maintenance Work

- Check gear box oil qty periodically and completely drain out oil and replace afresh as per manufacture's recommendation.
- Check oil pump every day ,top up if necessary.
- Check every day chain alignments and must periodically 4 hrs once chain cleaned or removed impurities materials.

Main use of Screening Water Treatment

- To remove the suspended solids.
- To avoided the pump cloaking.
- To increase the Bacteria attachment in the FBBR system.
- To reduce the Sludge Quantity in filter press.
- To reduce the short-circuit in Electro Coagulation System.
- To avoided the solids cloaking in the air distribution system in Equalization tank.
- To increase the air volume in Equalization tank.

Project Brief of Screening Water Treatment Plant

Auto is engaged into printing of textile garments using pigments and reactive colours. The Effluent Treatment Plant is designed to treat 40m3 per day of the effluent to achieve the norms prescribed by the local governing bodies and BSR guidelines. The Factory presently produce only up to 10m3/Day Effluent. The ETP is designed for higher capacity considering the future expansion and addition of Washing Process.

Plan Capacity of Screening Water Treatment Plant

The Effluent screening water treatment Plant is designed to treat wastewater generated from the processing unit. The Effluent treatment plant is designed to treat 40m3/day of Washing & Printing combined effluent generated from the process house (Washing 15m3/Day & Printing 5m3/Day). The plant will be capable of operating at the flow rate of 2m3/hr. The Present Effluent generation is 10m3/Day and the Plant is operated for 10-12 Hours per Day with 50% flow rate.

Organisms Involved in Water Purification

Water purification using organisms is a great natural method to treat mainly used water.

The implementing of purification using organisms is known and works in practice for many years now, the principals of the method are natural and occur in nature all the time regardless of human intervention. But the use of this type of biological water treatment method is becoming more known and also more common due to the overall understanding that mankind needs to find more sustainable and ecological ways to live, treat our waste and co-exist with other systems on the planet.

The major microbial populations found in wastewater treatment systems are bacteria, protozoa, viruses, fungi, algae and helminthes. The presence of most of these organisms in water leads to spread of diseases1. The majority of waterborne microorganisms that cause human disease come from animal and human fecal wastes. These contain a wide variety of viruses, bacteria, and protozoa that may get washed into drinking water supplies or receiving water bodies2. Microbial pathogens are considered to be critical factors contributing to numerous waterborne outbreaks. Many microbial pathogens in wastewater can cause chronic diseases with costly long-term effects, such as degenerative heart disease and stomach ulcer. The density and diversity of these pollutants can vary depending on the intensity and prevalence of infection in the sewered community. The detection, isolation and identification of the different types of microbial pollutants in wastewater are always difficult, expensive and time consuming. To avoid this, indicator organisms are always used to determine the relative risk of the possible presence of a particular pathogen in wastewater.

Viruses are among the most important and potentially most hazardous pollutants in wastewater. They are generally more resistant to treatment, more infectious, more difficult to detect and require smaller doses to cause infections4,5. Because of the difficulty in detecting viruses, due to their low numbers, bacterial viruses (bacteriophages) have been examined for use in faecal pollution and the effectiveness of treatment processes to remove enteric viruses. Bacteria are the most common microbial pollutants in wastewater. They cause a wide range of infections, such as diarrhea, dysentery, skin and tissue infections, etc. Disease-causing bacteria found in water include different types of bacteria, such as *E. coli* O157:H7; *Listeria* sp., *Salmonella* sp., *Leptosporosis* sp., etc (CDC, 1997). The major pathogenic protozoans associated with wastewater are *Giardia* sp. and *Cryptosporidium* sp. They are more prevalent in wastewater than in any other environmental source.

Pollutants in Wastewater

Pathogens

Parasites, bacteria and viruses may be injurious to the health of people or livestock ingesting the polluted water. These pathogens may have originated from sewage or from domestic or wild bird or mammal feces. Pathogens may be killed by ingestion by larger organisms, oxidation, infection by phages or irradiation by ultraviolet sunlight unless that sunlight is blocked by plants or suspended solids.

Suspended Solids

Particles of soil or organic matter may be suspended in the water. Such materials may give the water a cloudy or turbid appearance. The anoxic decomposition of some organic materials may give rise to obnoxious or unpleasant smells as sulphur containing compounds are released.

Nutrients

Compounds containing nitrogen, potassium or phosphorus may encourage growth of aquatic plants and thus increase the available energy in the local food-web. this can lead to increased concentrations of suspended organic material. In some cases specific micro-nutrients may be required to allow the available nutrients to be fully utilised by living organisms. In other cases, the presence of specific chemical species may produce toxic effects limiting growth and abundance of living matter.

Metals

Many dissolved or suspended metal salts exert harmful effects in the environment sometimes at very low concentrations. Some aquatic plants are able to remove very low metal concentrations, with the metals ending up bound to clay or other mineral particles.

Roles and Dynamics of Microorganisms in Wastewater Treatment Systems

The major microbial populations found in wastewater treatment systems are bacteria, protozoa, viruses, fungi, algae and helminthes. The presence of most of these organisms in water leads to spread of diseases.

Bacteria

In wastewater treatment systems, bacteria play vital role in the conversion of organic matter present to less complex compounds. In terms of size, bacteria range from 0.2- 2.0 μm in diameter and are responsible for most of the wastewater treatment in septic tanks. Though not all bacteria are harmful, a number of them cause water-related diseases in human and animals. Some of these diseases include cholera, dysentery, typhoid fever, salmonellosis and gastroenteritis. Bacteria can be found entangled in flocs as in the case of activated sludge, where some play important role in biological treatment while some like the filamentous bacteria can cause serious problem in settling and foaming. Waterborne gastroenteritis of unknown cause is frequently reported, with susceptible agent being bacteria. Certain strains of *Pseudomonas* and *Escherichia coli* which may affect the newborn are potential sources of this disease. These strains of microbes have also been implicated in gastrointestinal disease outbreaks.

Bacteria are of the greatest numerical importance in wastewater treatment systems. The majority is facultative living in either the presence or absence of oxygen. Although both heterotrophic and autotrophic bacteria are found in wastewater treatment systems thepredominant ones are the heterotrophic bacteria. Generally, heterotrophic bacteria obtain their energy from the carbonaceous organic matter in wastewater effluent. The energy obtained is used for the synthesis of new cells and also for the release of energy through the conversion of organic matter and water. Some important bacteria genera that are found in wastewater treatment systems are *Achromobacter, Alcaligenes, Arthrobacter, Citromonas, Flavobacterium, Pseudomonas, Zoogloe* and *Acinetobacter*.

In wastewater treatment systems, bacteria are responsible for the stabilization of influent wastes. The majority of the bacteria are known to form floc particles. The floc particles are clusters of bacteria that break down waste. Also, the floc particles also serve as sites on which waste can be absorbed and broken down. Filamentous bacteria form, which trichomes or filaments provide a backbone for the floc particles, allowing the particles to grow in size and withstand the shearing action in the treatment process. When filamentous bacteria are present in excessive numbers or length, they often cause solid/liquid separation or settleability problems.

Furthermore, bacteria are the most common microbial pollutants in wastewater. The presence of pathogenic bacteria can be indicated using the tests for total and faecal coliforms. Conventionally, the detection of feacal coliform is generally accepted as a reliableindicator of faecal contamination. *Escherichia coli*is also regarded as agood and reliable indicator forfaecal pollution from animal and human sources since it is known not to last for long periods outside the faecal environment. The tests for total and faecal coliforms can be carried out, using either the traditional or enzymatic methods. Traditionally, the tests for total and faecal coliforms are carried out using the multiple-tube fermentation or by membrane filtration techniques. While the multiple-tube fermentation technique is used for medium or highly contaminated waters, the membrane filtration technique is used for low or very low contaminated waters.

Protozoa

Protozoa are microscopic, unicellular organisms that are also found in the wastewater treatment systems. They perform many beneficial functions in the treatment process, including the clarification of the secondary effluent through the removal of bacteria, flocculation of suspended material and as bioindicators of the health of the sludge. The protozoa that inhabit the wastewater treatment systems are capable of movement in at least one stage of their development. They are 10 times bigger than bacteria; they are unicellular organisms with membrane enclosed organelles.

Protozoa prey on pathogenic bacteria which make it have an advantage in wastewater. They can be classified into five groups depending on their mode of locomotion, which are the free swimming ciliates, crawling ciliates, stalked and sessile ciliates, flagellates and amoeboid.

In wastewater treatment systems, protozoa are useful biological indicators of the condition of the systems. Although some protozoa are able to survive up to 12 h in the absence of oxygen, they are generally known as obligate aerobes, hence prove to be excellent indicators of an aerobic environment. Additionally, they serve as indicators of a toxic environment and are capable of exhibiting greater sensitivity to toxicity than bacteria. An indication of possible toxicity in a treatment system is the absence of or a lack of mobility of protozoa. The presence of large numbers of highly evolved protozoa in the biological mass in a wastewater treatment system is indicated as a hallmark of a well-operated and stable system.

They can be placed into one of five groups, according to their means of locomotion. These groups are the free-swimming ciliates, crawling ciliates and stalked/sessile ciliates, flagellates and amoebae. The three types of ciliates are free-swimming ciliates, crawling ciliates and stalked ciliates. All of these three have short hair-like structures or cilia that beat in unison to produce water current for locomotion and capturing bacteria. The water current moves suspended bacteria into a mouth opening.

In the aeration tank of biological processes, a true trophic web is established. The biological system of these plants consists of populations in continuous competition with each other for food. The growth of decomposers, prevalently heterotrophic bacteria, depends on the quality and quantity of dissolved organic matter in the mixed liquor. For predators, on the other hand, growth depends on the available prey. Dispersed bacteria are thus food for heterotrophic flagellates and bacterivorous ciliates, which, in turn, become the prey of carnivorous organisms. The relationships of competition and predation create oscillations and successions of populations until dynamic stability is reached. This is strictly dependent on plant management choices based on design characteristics aimed at guaranteeing optimum efficiency.

Some ciliates, however, are predators of other ciliates or are omnivorous, feeding on a variety of organisms including small ciliates, flagellates and dispersed bacteria. All bacterivorous ciliates rely on ciliary currents to force suspended bacteria to the oral region. Ciliated protozoa are numerically the most common species of protozoa in activated sludge, but flagellated protozoa and amoebae may also be present. The species of ciliated protozoa most commonly observed in wastewater treatment processes include *Aspidiscacostata, Carchesiumpolypinum, Chilodonellauncinata, Operculariacoarcta, Opercularíamicrodiscum, Trachelophyllumpusillum, Vorticella convallaria* and *Vorticella microstoma.*

There is indication that the free-swimming ciliates such as *Litonotus* sp. and *Paramecium* sp., which have cilia on all their body surfaces are typically found suspended or swimming freely in the bulk solution. On the other hand, the crawling ciliates, such as *Aspidisca* sp. and *Euplotes* sp. possess cilia only on their ventral or belly surface where the mouth opening is located. The crawling ciliates are usually found on floc particles while the stalked ciliates, such as *Carchesium* sp. and *Vorticella* sp., possess their cilia around the mouth opening only and are attached to floc particles. They have enlarged anterior portion and a slender posterior portion. The beating of the cilia and the springing action of the stalk produce a water vortex that draws dispersed bacteria into the mouth opening.

In wastewater systems, two types of amoebae are predominant, naked, such as *Actinophyrs* sp., *Mayorella* sp. and *Thecamoeba* sp. and the shelled amoebae or testate amoebae, e.g. *Cyclopyxis* sp. The naked amoeba lack any protective covering while shelled amoebae possess a protective covering that consists of calcified material. flagellated protozoa are oval in shape and possess one or more whip-like flagella. In wastewater treatment systems, the flagellated protozoa are propelled through the system by the help of the flagella in a cork-screw pattern of locomotion.

Viruses

Viruses are also found in wastewaters, particularly human viruses that are excreted in large quantities in faeces. Viruses that are native to animals and plants exist in smaller quantities in wastewater, although bacterial viruses may also be present. They are the causative agents of some water-related infections in humans, such as gastrointestinal and respiratory infections, conjunctivitis and meningitis. It is reported that a majority of waterborne diseases due to unidentified sources were caused by enteric viruses. They are very notorious and persistent when present in wastewater and can remain a viable source of infection for months after their entry into the wastewater.

Fungi

Fungi are also part of the microorganisms found in wastewater treatment systems. Fungi are multicellular organisms that are also constituents of the activated sludge. Under certain environmental conditions in a mixed culture, they metabolize organic compounds and can successfully compete with bacteria. Also, a small number of fungi are capable of oxidizing ammonia to nitrite, and fewer still to nitrate. The most common sewage fungus organisms are *Sphaerotilus natans* and *Zoogloea* sp.

A number of filamentous fungi are found naturally in wastewater treatment systems as spores or vegetative cells, although they can also metabolize organic substances. A number of fungi species, such as *Aspergillus, Penicillium, Fusarium, Absidia*and a host of others have been implicated in the removal of carbon and nutrient sources in wastewater. Some fungi are also reported to have the ability to breakdown organic matter present in the sludge system. In a system with low pH, where bacterial growth is inhibited, the main role of the fungi is the breakdown down of organic matter. Additionally, some fungi use their fungal hyphae for trapping and adsorbing suspended solids to accomplish their energy and nutrient requirements. Some filamentous fungi have been reported to secrete some enzymes which help in the degradation of substrates during wastewater treatment.

Algae

Algae can be found in wastewater because they are able to use solar energy for photosynthesis as well as nitrogen and phosphorus for their growth leading to eutrophication. Some types of algae that can be found in wastewater include *Euglena* sp., *Chlamydomonas* sp., and *Oscillatoria* sp. Algae are significant organisms for biological purification of wastewater because they can be able to accumulate plant nutrients, heavy metals, and pesticides, organic and inorganic toxic substances. The use of microalgae in biological wastewater treatment has gained a lot of importance over the years.

High rate algal pond is shallow and equipped with mechanical aeration and mixing by means of paddle wheels, 90% of BOD and 80% of nitrogen and phosphorus are treated in high rate algal ponds. Construction and energy costs are highly lower and the land requirement is not up to that of facultative pond in constructed wetlands.

Helminth

Nematodes are aquatic animals present in fresh, brackish and salt waters and wet or humid soil worldwide. Freshwater nematodes can be present in sand filters and aerobic treatment plants. They are present in large numbers in secondary wastewater effluents, biofilters and biological contractors. Freshwater nematodes inhabit freshwater below the water table with species utilizing oxygen dissolved in the fresh water. Nematodes are part of the ecosystem, serving as food for small invertebrates. They crawl onto floc particles and move in whip-like fashion when in the free-living mode. They secrete a sticky substance to be able to anchor to a substrate (media), so that anchored nematodes can feed without interference from currents or turbulence. A lack of nematode activity can be one of the bio-indicators of a toxic condition that may be developing in the treatment process.

There are many species of parasitic worms that have human hosts. Some of them have the ability to cause serious illnesses. They can be grouped into three, the roundworms, the flatworms and the annelids. The flatworms are divided into two groups, the tapeworms, which posses segments on their body and the flukes, which have a single, flat and unsegmented body. A common characteristic of most helminthes is that they reproduce through eggs, although the eggs may differ in size and shape. In wastewater, most helminthes eggs are not always infectious. For the eggs to be infectious they need to be viable and larval development must occur. The larval development occurs at required levels of temperature and moisture. Helminth eggs can remain viable for 1-2 months in crops and it takes longer months in soil, fresh water and sewage and can take years in faeces and sludge. Helminthes eggs are inactivated at elevate temperatureised (above 40°C) and reduced moisture (below 5%), although these conditions are not readily achieved during wastewater treatment; In typical water treatment, helminthes eggs are removed by physical means, such as sedimentation, filtration or coagulation-flocculation.

Secondary Treatment

Secondary treatment of sewage and other wastewater is the stage of wastewater treatment designed to substantially degrade the biological content of the sewage. This usually uses biological processes. Municipal and industrial plants usually use aerobic biological processes.

Suspended processes, particularly activated sludge, are most common in medium to large-scale plants; fixed film methods such as roughing filters need less maintenance and control, and are more resilient, and are appropriate where cost and maintenance are major issues.

Anaerobic treatment is sometimes used, in the form of septic tanks and in biogas digesters. in the case of septic tanks the primary and secondary phases are combined in one unit. If biogas digesters are used for secondary treatment, the primary treatment phase is reduced or emitted (aiming to remove matter such as gravel and litter rather than sewage solids).

Brief Explanation

Once wastewater has gone through the Primary treatment stage the Effluent will undergo a secondary treatment in order to remove both small suspended solids and BOD_5 (five day biochemical oxygen demand) that pass through the primary treatment stage. All secondary treatment systems use a biological process to break down organic matter. Microorganisms are introduced to the wastewater and consume the organic matter, oxygen is delivered to the system ensuring microorganism survival. Oxygen delivery differs among the various systems. This biological process occurs naturally in nature, but is accelerated in secondary treatment systems. Typically 85% of BOD and suspended solids are removed during this process. Water exiting secondary treatment will still carry nitrogen, phosphorus, heavy metals, Pathogens, and bacteria. For further removal of pollutants the water is transported to a tertiary treatment system and disinfection. There are a variety of secondary treatment processes; the following are conventional processes used by treatment plants:

- Activated sludge,
- Trickling filter,
- Non-electric secondary filtration (FilterPod),
- Oxidation ponds.

There are pros and cons to each of these three processes. Operational and initial costs along with space are three factors that will often determine which technique is appropriate. Space is influenced by population size and cost of land. For example, oxidation ponds require large areas of land if land is costly or needed for housing oxidation ponds are not a likely option. Additionally wastewater treatment plants need to consider maintenance, reliability, and effectiveness of the system.

Process Upsets

Process upsets are temporary decreases in treatment plant performance caused by significant population change within the secondary treatment ecosystem. Conditions likely to create upsets include for example toxic chemicals and unusually high or low concentrations of organic waste BOD providing food for the bioreactor ecosystem.

Toxicity

Waste containing biocide concentrations exceeding the secondary treatment ecosystem tolerance level may kill a major fraction of one or more important ecosystem species. BOD reduction normally accomplished by that species temporarily ceases until other species reach a suitable population to utilize that food source, or the original population recovers as biocide concentrations decline.

Dilution

Waste containing unusually low BOD concentrations may fail to sustain the secondary treatment population required for normal waste concentrations. The reduced population surviving the starvation event may be unable to completely utilize available BOD when waste loads return to normal. Dilution may be caused by addition of large volumes of relatively uncontaminated water such as stormwater runoff into a combined sewer. Smaller sewage treatment plants may experience dilution from cooling water discharges, major plumbing leaks, firefighting, or draining large swimming pools.

A similar problem occurs as BOD concentrations drop when low flow increases waste residence time within the secondary treatment bioreactor. Secondary treatment ecosystems of college communities acclimated to waste loading fluctuations from student work/sleep cycles may have difficulty surviving school vacations. Secondary treatment systems accustomed to routine production cycles of industrial facilities may have difficulty surviving industrial plant shutdown. Populations of species feeding on incoming waste initially decline as concentration of those food sources decrease. Population decline continues as ecosystem predator populations compete for a declining population of lower trophic level organisms.

Peak Waste Load

High BOD concentrations initially exceed the ability of the secondary treatment ecosystem to utilize available food. Ecosystem populations of aerobic organisms increase until oxygen transfer limitations of the secondary treatment bioreactor are reached. Secondary treatment ecosystem populations may shift toward species with lower oxygen requirements, but failure of those species to use some food sources may produce higher effluent BOD concentrations. More extreme increases in BOD concentrations may drop oxygen concentrations before the secondary treatment ecosystem population can adjust, and cause an abrupt population decrease among important species. Normal BOD removal efficiency will not be restored until populations of aerobic species recover after oxygen concentrations rise to normal.

Design for Damage Control

Measures creating uniform wastewater loadings tend to reduce the probability of upsets. Fixed-film or attached growth secondary treatment bioreactors are similar to a plug flow reactor model circulating water over surfaces colonized by biofilm, while suspended-growth bioreactors resemble a continuous stirred-tank reactor keeping microorganisms suspended while water is being treated. Secondary treatment bioreactors may be followed by a physical phase separation to remove biological solids from the treated water. Upset duration of fixed film secondary treatment systems may be longer because of the time required to recolonize the treatment surfaces. Suspended growth ecosystems may be restored from a population reservoir. Activated sludge recycle systems provide an integrated reservoir if upset conditions are detected in time for corrective action. Sludge recycle may be temporarily turned off to prevent sludge washout during peak storm flows when dilution keeps BOD concentrations low. Suspended growth activated sludge systems can be operated in a smaller space than fixed-film trickling filter systems that treat the same amount of water; but fixed-film systems are better able to cope with drastic changes in the amount of biological material and can provide higher removal rates for organic material and suspended solids than suspended growth systems.

Wastewater flow variations may be reduced by limiting stormwater collection by the sewer system, and by requiring industrial facilities to discharge batch process wastes to the sewer over a time interval rather than immediately after creation. Discharge of appropriate organic industrial wastes may be timed to sustain the secondary treatment ecosystem through periods of low residential waste flow. Sewage treatment systems experiencing holiday waste load fluctuations may provide alternative food to sustain secondary treatment ecosystems through periods of reduced use. Small facilities may prepare a solution of soluble sugars. Others may find compatible agricultural wastes, or offer disposal incentives to septic tank pumpers during low use periods.

Process Types

Filter Beds (Oxidizing Beds)

In older plants and those receiving variable loadings, trickling filter beds are used where the settled sewage liquor is spread onto the surface of a bed made up of coke (carbonized coal), limestone chips or specially fabricated plastic media. Such media must have large surface areas to support the biofilms that form. The liquor is typically distributed through perforated spray arms. The distributed liquor trickles through the bed and is collected in drains at the base. These drains also provide a source of air which percolates up through the bed, keeping it aerobic. Biofilms of bacteria, protozoa and fungi form on the media's surfaces and eat or otherwise reduce the organic content. The filter removes a small percentage of the suspended organic matter, while the majority of the organic matter supports microorganism reproduction and cell growth from the biological oxidation and nitrification taking place in the filter. With this aerobic oxidation and nitrification, the organic solids are converted into biofilm grazed by insect larvae, snails, and worms which help maintain an optimal thickness. Overloading of beds may increase biofilm thickness leading to anaerobic conditions and possible bioclogging of the filter media and ponding on the surface.

Rotating Biological Contactors

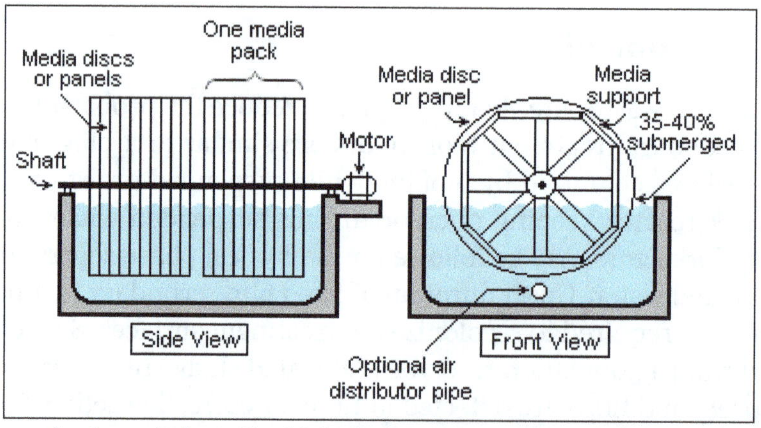

Schematic of a typical rotating biological contactor (RBC). The treated effluent clarifier/settler is not included in the diagram.

Rotating biological contactors (RBCs) are robust mechanical fixed-film secondary treatment systems capable of withstanding surges in organic load. RBCs were first installed in Germany in 1960 and have since been developed and refined into a reliable operating unit. The rotating disks support the growth of bacteria and micro-organisms present in the sewage, which break down and stabilize organic pollutants. To be successful, micro-organisms need both oxygen to live and food to grow. Oxygen is obtained from the atmosphere as the disks rotate. As the micro-organisms grow, they build up on the media until they are sloughed off due to shear forces provided by the rotating discs in the sewage. Effluent from the RBC is then passed through a secondary clarifier where the sloughed biological solids in suspension settle as a sludge.

Activated Sludge

Activated sludge is a common suspended-growth method of secondary treatment. Activated sludge plants encompass a variety of mechanisms and processes using dissolved oxygen to promote growth of biological floc that substantially removes organic material.[:12–13] Biological floc is an ecosystem of living biota subsisting on nutrients from the inflowing primary clarifier effluent. These mostly carbonaceous dissolved solids undergo aeration to be broken down and either biologically oxidized to carbon dioxide or converted to additional biological floc of reproducing micro-organisms. Nitrogenous dissolved solids (amino acids, ammonia, etc.) are similarly converted to biological floc or oxidized by the floc to nitrites, nitrates, and, in some processes, to nitrogen gas through denitrification. While denitrification is encouraged in some treatment processes, denitrification often impairs the settling of the floc causing poor quality effluent in many suspended aeration plants. Overflow from the activated sludge mixing chamber is sent to a secondary clarifier where the suspended biological floc settles out while the treated water moves into tertiary treatment or disinfection. Settled floc is returned to the mixing basin to continue growing in primary effluent. Like most ecosystems, population changes among activated sludge biota can reduce treatment efficiency. Nocardia, a floating brown foam sometimes misidentified as *sewage fungus*, is the best known of many different fungi and protists that can overpopulate the floc and cause process upsets. Elevated concentrations of toxic wastes including pesticides, industrial metal plating waste, or extreme pH, can kill the biota of an activated sludge reactor ecosystem.

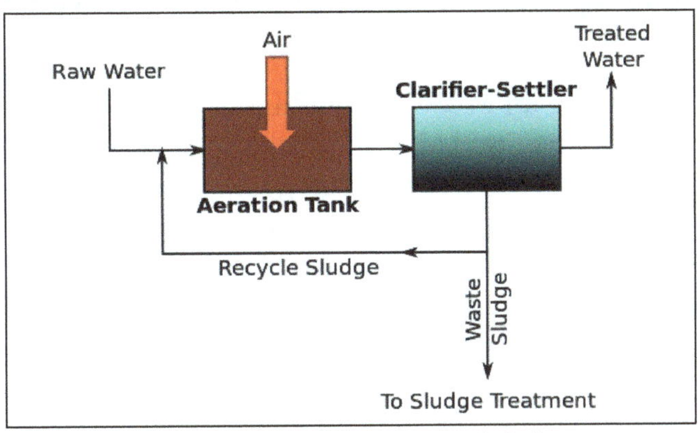

A generalized schematic of an activated sludge process.

Package Plants and Sequencing Batch Reactors

One type of system that combines secondary treatment and settlement is the cyclic activated sludge (CASSBR), or sequencing batch reactor (SBR). Typically, activated sludge is mixed with raw incoming sewage, and then mixed and aerated. The settled sludge is run off and re-aerated before a proportion is returned to the headworks.

The disadvantage of the CASSBR process is that it requires a precise control of timing, mixing and aeration. This precision is typically achieved with computer controls linked to sensors. Such a complex, fragile system is unsuited to places where controls may be unreliable, poorly maintained, or where the power supply may be intermittent. Extended aeration package plants use separate basins for aeration and settling, and are somewhat larger than SBR plants with reduced timing sensitivity.

Package plants may be referred to as *high charged* or *low charged*. This refers to the way the biological load is processed. In high charged systems, the biological stage is presented with a high organic load and the combined floc and organic material is then oxygenated for a few hours before being charged again with a new load. In the low charged system the biological stage contains a low organic load and is combined with flocculate for longer times.

Membrane Bioreactors

Membrane bioreactors (MBR) are activated sludge systems using a membrane liquid-solid phase separation process. The membrane component uses low pressure microfiltration or ultrafiltration membranes and eliminates the need for a secondary clarifier or filtration. The membranes are typically immersed in the aeration tank; however, some applications utilize a separate membrane tank. One of the key benefits of an MBR system is that it effectively overcomes the limitations associated with poor settling of sludge in conventional activated sludge (CAS) processes. The technology permits bioreactor operation with considerably higher mixed liquor suspended solids (MLSS) concentration than CAS systems, which are limited by sludge settling. The process is typically operated at MLSS in the range of 8,000–12,000 mg/L, while CAS are operated in the range of 2,000–3,000 mg/L. The elevated biomass concentration in the MBR process allows for very effective removal of both soluble and particulate biodegradable materials at higher loading rates. Thus increased sludge retention times, usually exceeding 15 days, ensure complete nitrification even in extremely cold weather.

The cost of building and operating an MBR is often higher than conventional methods of sewage treatment. Membrane filters can be blinded with grease or abraded by suspended grit and lack a clarifier's flexibility to pass peak flows. The technology has become increasingly popular for reliably pretreated waste streams and has gained wider acceptance where infiltration and inflow have been controlled, however, and the life-cycle costs have been steadily decreasing. The small footprint of MBR systems, and the high quality effluent produced, make them particularly useful for water reuse applications.

Aerobic Granulation

Aerobic granular sludge can be formed by applying specific process conditions that favour slow growing organisms such as PAOs (polyphosphate accumulating organisms) and GAOs (glycogen accumulating organisms). Another key part of granulation is selective wasting whereby slow settling floc-like sludge is discharged as waste sludge and faster settling biomass is retained. This process has been commercialized as Nereda process.

Surface-aerated Lagoons or Ponds

Aerated lagoons are a low technology suspended-growth method of secondary treatment using motor-driven aerators floating on the water surface to increase atmospheric oxygen transfer to the lagoon and to mix the lagoon contents. The floating surface aerators are typically rated to deliver the amount of air equivalent to 1.8 to 2.7 kg O_2/kW·h. Aerated lagoons provide less effective mixing than conventional activated sludge systems and do not achieve the same performance level. The basins may range in depth from 1.5 to 5.0 metres. Surface-aerated basins achieve 80 to 90 percent removal of BOD with retention times of 1 to 10 days. Many small municipal sewage systems in the United States (1 million gal./day or less) use aerated lagoons.

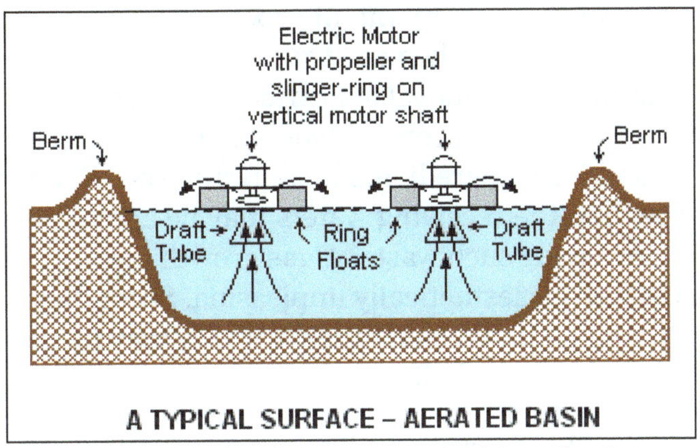

A typical surface-aerated basin (using motor-driven floating aerators).

Constructed Wetlands

Primary clarifier effluent was discharged directly to eutrophic natural wetlands for decades before environmental regulations discouraged the practice. Where adequate land is available, stabilization ponds with constructed wetland ecosystems can be built to perform secondary treatment separated from the natural wetlands receiving secondary treated sewage. Constructed wetlands resemble fixed-film systems more than suspended growth systems, because natural mixing is minimal. Constructed wetland design uses plug flow assumptions to compute the residence time required for treatment. Patterns of vegetation growth and solids deposition in wetland ecosystems, however, can create preferential flow pathways which may reduce average residence time. Measurement of wetland treatment efficiency is complicated because most traditional water quality measurements cannot differentiate between sewage pollutants and biological productivity of the wetland. Demonstration of treatment efficiency may require more expensive analyses.

Emerging Technologies

- Biological Aerated (or Anoxic) Filter (BAF) or Biofilters combine filtration with biological carbon reduction, nitrification or denitrification. BAF usually includes a reactor filled with a filter media. The media is either in suspension or supported by a gravel layer at the foot of the filter. The dual purpose of this media is to support highly active biomass that is attached to it and to filter suspended solids. Carbon reduction and ammonia conversion occurs in aerobic mode and sometime achieved in a single reactor while nitrate conversion occurs in anoxic mode. BAF is operated either in upflow or downflow configuration depending on design specified by manufacturer.

- Integrated Fixed-Film Activated Sludge.
- Moving Bed Biofilm Reactors typically requires smaller footprint than suspended-growth systems.

Tertiary Wastewater Treatment

Tertiary treatment is the next wastewater treatment process after secondary treatment. This treatment is sometimes called as the final or advanced treatment and consists of removing the organic load left after secondary treatment for removal of nutrients from sewage and particularly to kill the pathogenic bacteria. The effluents from secondary sewage treatment plants contain both nitrogen (N) and phosphorus (P). N and P are ingredients in all fertilizers. When excess amounts of N and P are discharged, plant growth in the receiving waters may be accelerated which results in eutrophication in the water body receiving such waste. Algae growth may be stimulated causing blooms which are toxic to fish life as well as aesthetically unpleasing. Secondary treated effluent also contains suspended, dissolved, and colloidal constituents which may be required to be removed for stipulated reuse or disposal of the treated effluent.

The purpose of tertiary treatment is to provide a final treatment stage to raise the effluent quality before it is discharged to the receiving environment such as sea, river, lake, ground, etc., or to raise the treated water quality to such a level to make it suitable for intended reuse. This step removes different types of pollutants such as organic matter, SS, nutrients, pathogens, and heavy metals that secondary treatment is not able to remove. Wastewater effluent becomes even cleaner in this treatment process through the use of stronger and more advanced treatment systems. It includes sedimentation, coagulations, membrane processes, filtration, ion exchange, activated carbon adsorption, electrodialysis, nitrification and denitrification, etc. Tertiary treatment is costly as compared to primary and secondary treatment methods.

Need of Tertiary Treatment

- Tertiary treatment may be provided to the secondary effluent for one or more of the following contaminant further.
- To remove total suspended solids and organic matter those are present in effluents after secondary treatment.
- To remove specific organic and inorganic constituents from industrial effluent to make it suitable for reuse.

- To make treated wastewater suitable for land application purpose or directly discharge it into the water bodies like rivers, lakes, etc.
- To remove residual nutrients beyond what can be accomplished by earlier treatment methods.
- To remove pathogens from the secondary treated effluents.
- To reduce total dissolved solids (TDS) from the secondary treated effluent to meet reuse quality standards.

One or more of the unit operation/ process mentioned in Figure will be used for achieving this tertiary treatment.

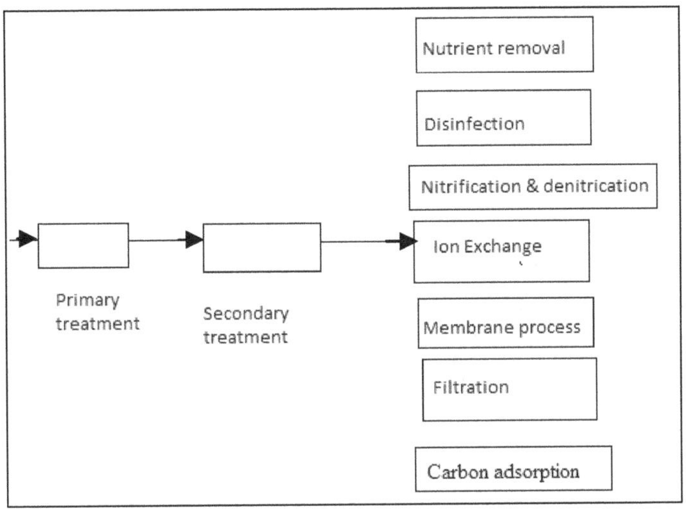

Process involved in tertiary wastewater treatment.

Tertiary Treatments

In advanced wastewater treatment, treatment options or methods are dependent upon the characteristics of effluent to be obtained after secondary treatment to satisfy further use or disposal of treated wastewater.

Nitrogen Removal

Wastewater containing nutrients includes sewage, agriculture runoff and many of the industrial effluents. The nutrients of most concerned are N and P. The principal nitrogen compounds in domestic sewage are proteins, amines, amino acids, and urea. Ammonia nitrogen in sewage results from the bacterial decomposition of these organic constituents.

The nitrogen compounds results from the biological decomposition of proteins and from urea discharged in body waste. This nitrogen is in complex organic molecules and is referred simply as organic nitrogen. Organic nitrogen may be biologically converted to free ammonia (NH_3^o) or to the ammonium ion (NH_4^+) by one of several different metabolic pathways. These two exists in equilibrium as,

$$NH_4^+ \rightleftharpoons NH_3 + H^+$$

Ammonia nitrogen is the most reduced nitrogen compound found in wastewater, which can be biologically oxidized to nitrate if molecular oxygen is present (under aerobic condition). In wastewater, the predominant forms of nitrogen are organic nitrogen and ammonia. The nitrification may takes place in biological treatment units provided the treatment periods are long enough. Generally, for the HRT used in secondary treatment conversion of organic nitrogen to ammonia is significant and nitrification may not be significant. Because of oxygen demand exerted by ammonia (about 4.6 mg of O_2 per mg of NH^+-N oxidized) and due to other environmental factors, removal of ammonia may be required. The most common processes for removal of ammonia from wastewater are:

i) Air stripping,

ii) Biological nitrification and denitrification.

Air Stripping

It consists of converting ammonium to gaseous phase and then dispersing the liquid in air, thus allowing transfer of the ammonia from wastewater to the air. The gaseous phase NH_3^o and aqueous phase NH_4^+ exist together in equilibrium. The relative abundance of these phases depends upon both the pH and the temperature of the wastewater. The pH must be greater than 11 for complete conversion to NH_3. Since, this pH is greater than pH of normal wastewater, adjustment of pH is necessary prior to air stripping. Addition of lime is the most common means for raising the pH. Enough lime must be added to precipitate the alkalinity and to add the excess OH^- ions for pH adjustment.

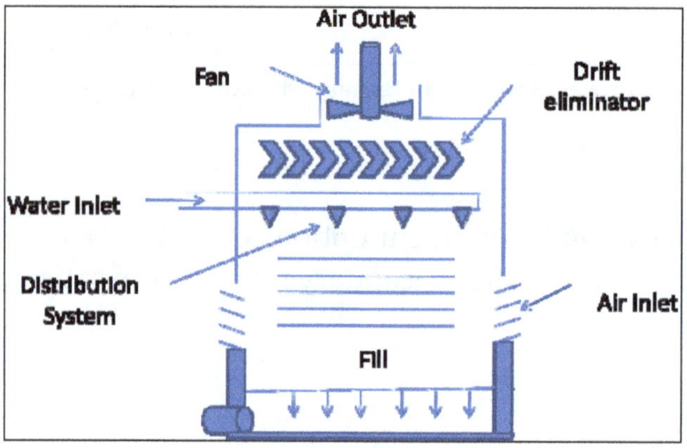

Spray tower

The most important and efficient reactor for air stripping is counter current spray tower. Larger quantity of air is required, and fan must be installed to draw air through tower. Packing is provided to minimize the film resistance to gas transfer by continuously forming, splashing and reforming drops.

The air to wastewater ratio ranging from 2000 to 6000 m^3 of air/m^3 of wastewater is used for design. Air requirement is more at lower temperature. Tower depths are generally less than 7.5 m, and hydraulic loading vary from 40 to 46 L/min.m^2 of tower. The limitation to this process occurs at temperature close to freezing temperature. Drastic reduction in efficiency is observed and preheating of gas is not possible because of high volume.

The other problems associated with this include noise and air pollution and scaling of the packing media. Noise pollution problem is caused by roar of the fan. The odor problem is due to dispersion of ammonia gas in atmosphere, due to which this technique may not be permitted at some location. This problem can be minimized by locating the facility away from the populated area. Precipitation of calcium carbonate scale on the packing media as a result of wastewater softening can be minimized by the use of smooth surface PVC pipe as packing material. The occasional cleaning of packing media is still required.

Biological Nitrification and Denitrification

Bacteria remove ammonia nitrogen from wastewater by a two step biological processes: nitrification followed by denitrification to covert it finally to gaseous nitrogen. In this gaseous form N2 is inert and does not react with the wastewater itself or with other constituents present in wastewater. Since, treated wastewater is likely to be saturated with molecular nitrogen; the produced N2 is simply released to the atmosphere. These two steps involved require different environmental conditions and hence generally they are carried out in separate reactors.

Nitrification

It has important role in nitrogen removal from wastewater during treatment. The biological conversion of ammonium to nitrate nitrogen is called Nitrification. It is autotrophic process i.e. energy for bacterial growth is derived by oxidation of nitrogen compounds such as ammonia. In this process, the cell yield per unit substrate removal is smaller than heterotrophs. Nitrification is a two-step process. In first step, bacteria known as *Nitrosomonas* can convert ammonia and ammonium to nitrite. These bacteria known as nitrifiers are strictly aerobes. This process is limited by the relatively slow growth rate of *Nitrosomonas*. Next, bacteria called *Nitrobacter* finish the conversion of nitrite to nitrate.

Nitrosomonas

First step: $NH_4^+ + 3/2\ O_2 \longrightarrow No_2^- + 2H^+ + H_2O$ (Nitrite)

Nitrobacter

Second step: $No_2^- + 1/2\ O_2 \longrightarrow$ (Nitrate)

Overall reaction: $NH_4^+ + 2\ O_2 \longrightarrow No_3^- + 2H^+ + H_2O$

Nitrosomonas and *Nitrobacter* use the energy derived from the reactions for cell growth and maintenance. Some of ammonium ions are assimilated into cell tissues. Neglecting this ammonium ion used in cell synthesis the O_2 required to oxidize ammonia to nitrate is 4.57 mg O_2/mg ammonium nitrogen. If the ammonium used in cell, O2 required is considered it is 4.3 mg O_2/mg ammonium nitrogen and about 7.14 mg of alkalinity is needed to neutralize the H^+ produced.

Nitrification may be used to prevent oxygen depletion from nitrogenous demand in the receiving water. Nitrification requires a long retention time, a low food to microorganism ratio (F/M), a high mean cell residence time (MCRT), and adequate alkalinity. Wastewater temperature and Ph affects the rate of nitrification.

Under favourable conditions, carbon oxidation and nitrification may occur in a single reactor called single stage under favourable conditions. In separate stage carbon oxidation and nitrification occur in different reactors. It can be accomplished in both suspended and attached growth process such as trickling filter, ASP, rotating disc biological contactor (RBC), SBR, etc. Also, nitrifying organisms are present in almost all aerobic process sludge but they are less in number. In nitrification, when ratio of BOD_5 to TKN is greater than 5, the process is called as combined carbon oxidation and nitrification, whereas, when ratio of BOD_5 to TKN is less than 3, the process is called as separate stage carbon oxidation and nitrification.

For single stage carbon oxidation and nitrification, plug flow, completely mixed and extended aeration ASP can be used. Oxidation ditch is also one of option for this process. The attached growth processes like TF and RBC are commonly used. Nitrification can be achieved by reducing applied loading rate, increasing the mean cell residence time (θ_c) and air supply.

Nitrifying bacteria are sensitive organisms. A variety of organic and inorganic agents can inhibit the growth and action of these organisms. High concentration of ammonia and nitrous acid can be inhibitory. The effect of pH is also significant with optimal range of 7.5 to 8.6. The system acclimatize to lower pH can also work successfully. The temperature also has considerable impact on growth of the nitrifying bacteria. Dissolved oxygen concentration above 1 mg/L is essential for nitrification. Below this DO, oxygen becomes the limiting nutrients and nitrification slows down or ceases.

Denitrification

In some applications, such as discharge of effluent into enclosed water bodies or recycle to water supplies, nitrification may not be sufficient. When nitrogen removal is required, one of the available methods is to follow biological nitrification with denitrification. Denitrification is accomplished under anaerobic or near anaerobic conditions by facultative heterotrophic bacteria commonly found in wastewater. Nitrates are removed by two mechanisms: (1) conversion of NO_3 to N_2 gas by bacterial metabolism and (2) conversion of NO_3 to nitrogen contained in cell mass which may be removed by settling. Denitrification occurs when oxygen levels are depleted and nitrate becomes the primary electron acceptor source for microorganisms.

$$\text{Nitrate}, NO_3^- \rightarrow \text{Nitrate}, NO_2^- \rightarrow \text{Nitrate oxide}, NO \rightarrow \text{Nitrous oxide}, N_2O \rightarrow \text{Nitrogen}, N_2$$

Denitrifying bacteria are facultative organisms, they can use either dissolved oxygen or nitrate as an oxygen source for metabolism and oxidation of organic matter. This is carried out by hetetrophic bacteria such as *pseudomonas, spirillum, lactobacillus, bacillus, microaoccus*, etc. For reduction to occur, the DO level must be near to zero, and carbon supply must be available to the bacteria. Because of low carbon content is required for the previous nitrification step, carbon must be added before denitrification can proceed. A small amount of primary effluent, bypassed around secondary and nitrification reactor can be used to supply the carbon. However, the unnitrified compounds in this water will be unaffected by the denitrification process and will appear in effluent.

When complete nitrogen removal is required, an external source of carbon containing no nitrogen will be required. The most commonly used external source of nitrogen is methanol. When methanol is added the reaction is,

$$NO_3^- + 5/6CH_3OH \rightarrow \tfrac{1}{2}N_2 + 5/6CO_2 + 7/6H_2O + OH^-$$

For treatment plant above 3 mg/L of methanol is required for each milligram per litre of nitrate, making this process an expensive. Alkalinity is generated in this process. Denitrification can be carried out as attached growth (anaerobic filter) and suspended growth process (expanded bed or UASB reactor).

Phosphorus Removal

Normally secondary treatment can only remove 1-2 mg/l of phosphorus, so a large excess of phosphorus is discharged in the final effluent, causing eutrophication of lakes and natural water bodies. Generally it appears as orthophosphate, polyphosphate and organically bound phosphorus. Phosphorus is required for microbes for cell synthesis and energy transport, maintenance as well as it is stored for subsequent use by microbes. During secondary treatment process about 10 to 30 % of influent phosphorus is removed (Metcalf & Eddy, 2003). Phosphate removal is currently achieved largely by chemical precipitation, which is expensive and causes an increase of sludge volume by up to 40%. An alternative is the biological phosphate removal (BPR), which is accomplished by sequencing and producing the appropriate environmental condition in the reactors.

Acinetobacter organisms are helpful for removal of phosphorus from effluent. Under anaerobic conditions, phosphorus accumulating organisms (PAO) assimilate fermentation products (i.e. volatile fatty acids) into storage products within the cells with the concomitant release of phosphorus from stored polyphosphates (Gray, 2005). Release of phosphorus occurs under anoxic condition. The BPR requires both aerobic and anaerobic zones in reactors for efficient treatment. Generally, lime precipitation is most commonly used for phosphorus removal. Phosphorus is removed in the waste sludge from the system

Treatment technologies presently available for phosphorus removal include (de-Bashan and Bashan, 2004):

A) Physical:

- a) Filtration for particulate phosphorus.
- b) Membrane technologies.

B) Chemical:

- a) Precipitation.
- b) Other (mainly physical-chemical adsorption).

C) Biological:

- a) Assimilation.
- b) Enhanced biological phosphorus removal (EBPR).

Ion Exchange

Ion Exchange can be used in wastewater treatment plants to swap one ion for another for the purpose of demineralization. The widest application of this process is in domestic water softening, where sodium ions are removed on cation exchange resin and chlorides are removed on anion exchange resin. Ion exchange is a unit process in which ions are removed out of an aqueous solution and is replaced by another ionic species. The basic principle behind ion exchange is that a weakly bound ion can preferably be displaced by a stronger binding ion. This effect is called the principle of selectivity. A more selective ion binds more strongly than a less selective ion. The effect of selectivity can be used to remove distinct ions from water and to replace them with others. It can be operated in a batch or continuous mode and has been used for removal of nitrogen, heavy metals and TDS in wastewater applications. It has also been used selectively to remove specific impurities and to recover valuable trace metals like chromium, nickel, copper, lead and cadmium from industrial waste discharges.

A number of naturally occurring minerals have ion exchange properties. Among them the notable ones are aluminium silicate minerals, which are called zeolites. Synthetic zeolites have been prepared using solutions of sodium silicate and sodium aluminate. Alternatively synthetic ion-exchange resins composed of organic polymer with attached functional groups such as – SO_3^- H^+ (strongly acidic cation exchange resins), or - COO^- - H^+ (weakly acidic cation exchange resins or -$N^+(CH_3)_3OH^-$ (strongly basic anion exchange resins) can be used. Synthetic and industrially produced ion exchange resins consist of small, porous beads that are insoluble in water and organic solvents. The most widely used base-materials are polystyrene and polyacrylate.

Membrane Process

Membrane technology can be used to treat a variety of wastes, including sanitary landfill leachate containing both organic and inorganic chemical species, water-soluble oil wastes used in metal fabricating and manufacturing industries, solvent-water mixtures, and oil-water mixtures generated during washing operations at metal fabricating facilities. Depending upon the material used for membrane, nature of driving force and separation mechanism, the membrane processes can be classified into sub-processes such as electrodialysis (ED) or electrodialysis reversal (EDR), microfiltration (MF), ultrafiltration (UF), nanofiltration (NF), reverse osmosis (RO).

MF, UF, NF and RO use pressure to transport water across the membrane. MF removes particulate matter while RO removes many solutes as water permeate through the membrane. MF membranes have the largest pore size and typically reject large particles and various microorganisms. UF membranes have smaller pores than MF membranes and, therefore, in addition to large particles and microorganisms, they can reject bacteria and soluble macromolecules such as proteins. RO membranes are effectively non-porous and, therefore, exclude particles and even many low molar mass species such as salt ions, organics, etc. NF membranes are relatively new and are sometimes called loose RO membranes. They are porous membranes, but since the pores are of the order of ten angstroms or less, they exhibit performance between that of RO and UF membranes. Different membrane processes are being used in water treatment since 1960.

The driving forces for the use of membrane technology are increased regulatory pressure to provide better quality water, increased demand of water requiring exploitation of low quality water resources as source and development and commercialization of membrane processes. Application of membrane processes for wastewater treatment is increasing worldwide day by day with reduction in cost of process and increased water charges. Membrane processes are used for about 70 % of the total installed capacity of desalination worldwide and percentage is increasing (for drinking water). In Japan, more than 30 MLD wastewater is used as toilet flushing water in building. It has application in many industrial wastewater treatment plants to produce reusable quality water from the effluents.

Types of Membrane Operation

The membrane is defined as the thin film separating two phases and acting as a selective barrier to the transport matter. Chemical potential difference exists between the two phases. *Retentate* contains non permeating species. *Permeate* forms the produced water from the membrane filtration. Membrane operation is recommended nomenclature than membrane process. The relevant main membrane operations used in water treatment are summarized in Table.

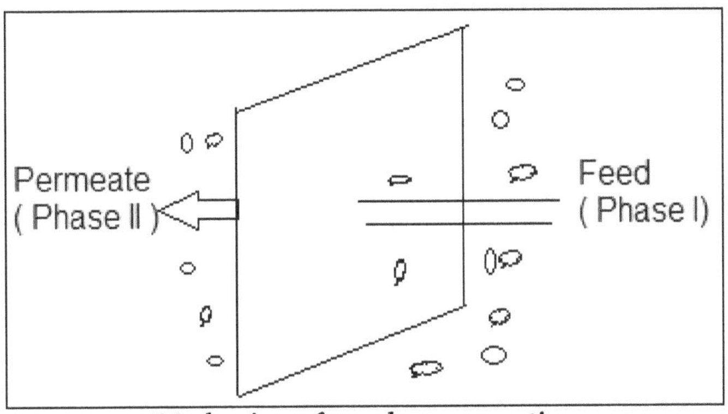

Mechanism of membrane operation.

Table: Technically relevant main membrane operations in water treatment.

Membrane	Driving force	Mechanism of separation	Membrane structure	Phases in contact
Microfiltration (MF)	pressure	Sieve	Macropores	Liquid-liquid
Ultrafiltration (UF)	pressure	Sieve	Mesopores	Liquid-liquid
Nanofiltration (NF)	pressure	Sieve + solution + diffusion + exclusion	Micropores	Liquid-liquid
Reverse Osmosis	pressure	Solution/diffusion + exclusion	Dense	Liquid-liquid

Advantages of membrane operation:

- Separation takes place at ambient temperature without phase change.
- Separation takes place without accumulation of product inside membrane (unlike ion exchange resins, which needs replacement/regulations).
- No need of chemical additives for separations.

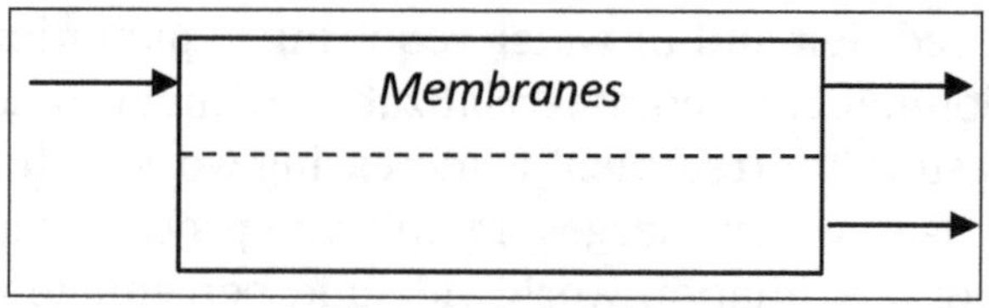

Three end modules

Membrane Operations

The transmembrane flux for each element is given by

Flux = Force * concentration * mobility

In most cases, concentration will vary with distance from the membrane surface along the boundary layer. Hence above equation is local equation where local forces are used to estimate gradient of chemical potential dμ/dx of every component that can be transported.

The variation of chemical potential of component 'i' can be expressed as a sum of the terms,

$$d\mu_i = \mathrm{RTd}\,\ln a_i + V_i\ dP + Z_i\ Fd\Psi$$

where, activity a_i (product of concentration by activity coefficient) is not under the control of operator. The pressure P and electric potential Ψ can be varied in order to improve the separation between the mobile components. The applied pressure of every component is proportional to its molar volume V_i. Electric field acts on every ionic species according to its valency Z_i and does not affect non-ionic species.

Reverse Osmosis

Reverse osmosis (RO) is a most commonly used membrane filtration method that removes many types of large molecules and ions from effluents by applying pressure to the effluents when it is on one side of a selective membrane. RO is used to remove specific dissolved organic constituents from wastewater remaining after advanced treatment with depth filtration or MF. RO system can operate at 90 % efficiency or better with respect to TDS. In addition, it also removes residual organic molecules, turbidity, bacteria and viruses.

The solvent of the solution is transferred through a dense membrane tailored to retain salts and low molecules weight solutes. When concentrated saline solution is separated from pure water by such a membrane, difference in chemical potential tends to promote the diffusion of water from the diluted compartment to the concentrated compartment in order to equalize concentration. At equilibrium, the difference in the levels between two compartments corresponds to the osmotic pressure of the saline solution. The demineralization of water can be accomplished using thin, microporous membranes. There are two basic modes of operation in use. One system uses pressure to drive water through the membrane against the force of osmotic pressure and is called RO. The pressure applied is several orders of magnitude in excess of natural osmotic pressure. The membrane commonly used in reverse osmosis is composed of cellulose acetate and is about 100 μm thick. The film contains microscopic openings that allow water molecules to pass through but retain dissolved solids by either molecular sieving (or by some other mechanism which is not yet completely understood). The process results in a concentrated solution of the ions on the pressure side of the membrane and product water relatively free of ions on the other side of the membrane.

For the producing pure water from the saline solution, osmotic pressure of the solution must be exceeded in the brine. For economically viable flows, at least twice the osmotic pressure must be exerted e.g. for seawater pressure of 5 -8 MPa is used in practice.

Nanofiltration (NF)

Nanofiltration (NF) is a rapidly advancing membrane separation technique for water and wastewater treatment as well as concentration/separation of antibiotics and pharmaceuticals due to its unique charge-based repulsion property and high rate of permeation. This process is also called as low pressure RO or membrane softening. It lies between RO and ultrafiltration in terms of selectivity of membranes, which is designed for removal of multivalent ions (Ca, Mg, etc.) in softening operation. Monovalent ions are poorly rejected. Hence, osmotic back pressure is much lower than that in RO. The operating pressure used in NF is typically 0.5 to 1.5 MPa.

It is a pressure driven process wherein the pore size of the membrane is in the range of 0.5-1 nm. Due to the lower operating pressure and higher flow rates, nanofiltration is inexpensive when compared to reverse osmosis. NF membranes allow partial permeation of monovalent salts such as sodium chloride while rejecting bivalent salts and hardness to a greater extent from aqueous solutions. NF can lower TDS and hardness, reduce color and odor, and remove heavy metal ions from groundwater. Other possible applications include treatment of effluents from textile dyeing, bulk drug and chemical process industries.

NF process is useful for raw water containing TDS in the range 700-1200 ppm and has exhibited two advantages over RO. NF allows sufficient permeation of essential salts while keeping the total dissolved solids (TDS) in accordance to BIS drinking water standards, whereas RO removes almost all the minerals essential for the human body. NF can be operated at lower pressure while RO requires higher operating pressure and therefore higher running cost.

Ultrafiltration (UF)

It is a clarification and disinfection membrane operation. UF membranes are porous and allow only coarser solutes (macromolecules) to be rejected. All types of microorganisms as viruses and bacteria and all types of particles can be removed by this process. Since, the low molecular solutes are not retained by UF, osmotic back pressure can be neglected and operating pressure is kept low as 50- 500 KPa.

Microfiltration (MF)

Microfiltration (MF) membranes are having 0.1 μm or more pore size. It is generally used for particulate matter removal. The pressure used in this process is similar to that of UF.

Classification of Membranes

According to separation mechanism: Three mechanisms exist for separation of ions and other particulate matter from the membranes, and accordingly the membranes are classified as below:

- Sieve effect: In this mechanism, separation is based on difference in pore size e.g. MF, UF.
- Solution-diffusion mechanism: In this mechanism, separation is based on difference in the solubility and diffusivity of materials in the membrane e.g. RO.
- Electrochemical effect: In this mechanism, separation is based on difference in the charges of the species to be separated e.g. ED (electrodialysis).

The classification based on separation mechanism leads to three main classes of membranes.

i) Porous membranes

Fixed pores are present in these membranes. These pores are sub-divided into three types' viz.

- Macropores: Theses are larger than 50 nm.
- Mesopores- These are in the range of 20 to 50 nm.
- Micropores- These are in the size less than 2 nm.

MF and UF are porous membranes while, NF could be classified in an intermediate class between porous and non porous membranes. Since, solution diffusion and even electrochemical effect have to introduce in equilibrium of mass transfer.

ii) Non Porous membranes

These are dense media membranes. The diffusion of species takes place in the free volume which is present between the macromolecular chains of the membrane material. RO is non porous membrane.

iii) Ion exchange membranes

These are specific types of non porous membranes. They consist of highly swollen gels carrying fixed positive or negative charges. A membrane with fixed positive charges (e.g. $-NR_3^+$) is called as anion exchange membranes, whereas cation exchange membranes have fixed negative charges (e.g. $-SO^{3-}$).

Classification based on Morphology

i) Asymmetric membranes

These are anisotropic membranes prepared from the same material.

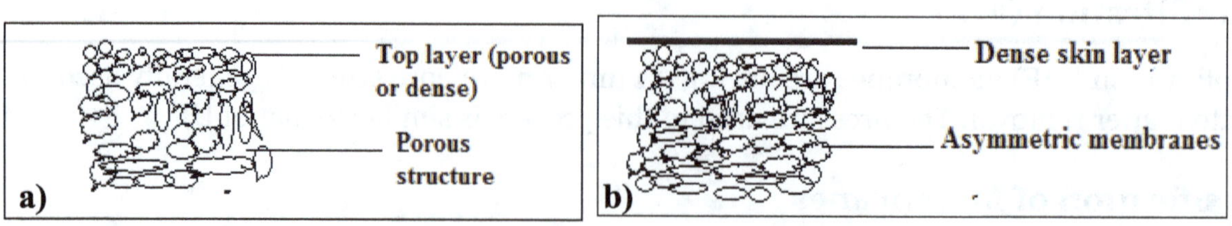

Asymmetric membrane and b) Composite membrane.

ii) Composite membranes

These are anisotropic membranes where top layer and sub-layer originate from different materials. Each layer can be optimized independently. Generally porous layer is asymmetric membranes.

The anisotropic membrane consists of very thin top layer called skin supported by a thicker and more porous supporting sub-layer. Skin has main function of membrane. Overall flux and selectivity depends upon the structure of skin. Skin has thickness in the range of 0.1 to 0.5 μm, which is about 1% of the thickness of porous sub-layer. Supporting layer offers negligible resistance to mass transfer and imparts mechanical strength to the membrane forming integral part of the membrane.

Classification based on Geometry

According to the geometry, the membranes can be classified as flat sheet membrane and cylindrical membrane. The cylindrical membrane can be further classified as (a) Tubular membranes having internal diameter >3 mm, and (b) Hollow fiber membranes having internal diameter < 3 mm. They are available in market with outer diameter ranging from 80 to 500 μm. They are used in RO with larger diameter. Also hallow fiber membranes are used in MF and UF in which they are also called as 'capillary membranes'.

Classification based on Chemical Nature

On the basis of chemical nature of the material used for membrane these are classified as: organic (polymers) membrane and inorganic membranes made from metals, ceramics, glasses, etc.

a) Polymer (organic) membranes

Cellulose and its derivatives are more widely used. These hydrophilic polymers are low cost and they are having low tendencies for adsorption. Polyamides, a hydrophilic polymers, are second type of polymer (aromatic polyamides) used for making membrane after cellulose diacetate. It has better thermal, chemical and hydraulic stability than cellulose esters. Amide group (-CO-NH-) cannot tolerate exposure to trace of Cl2, hence they are not suitable for chlorinated water. Poly acrylonitrile (PAN) are used in UF but not in RO.

Polysulphone (PSF) and Poly ethersulphone (PES) - These two are not hydrophilic and hence they have high adsorption tendency. They have very good thermal, chemical and mechanical stability. Polytetra fluoro ethylene (PTFE), Polyvinylidene fluoride (PVDF), Polyethylene (PE), Polycarbonate (PC) or isotactic Polypropylene (PP) are hydrophilic polymers that are used for membranes.

b) Inorganic membranes

These membranes have superior thermal, chemical and mechanical stability relative to polymer materials. The disadvantage of this membrane is that they are brittle and more expensive. Ceramic membranes are oxides, nitrides or carbides of metals such as Al, Zr, Ti, etc.

References

- Messina, Gabriele (October 2015). "A new UV-LED device for automatic disinfection of stethoscope membranes" (PDF). American Journal of Infection Control. Elsevier. Retrieved 11, January 2020
- "Ultraviolet light disinfection in the use of individual water purification devices" (PDF). U.S. Army Public Health Command. Retrieved 21, March 2020
- Tchobanoglous, G., Burton, F.L., and Stensel, H.D. (2003). Wastewater Engineering (Treatment Disposal Reuse) / Metcalf & Eddy, Inc (4th ed.). McGraw-Hill Book Company. ISBN 0-07-041878-0
- Ware, M. W.; et al. "Inactivation of Giardia muris by low pressure ultraviolet light" (PDF). United States Environmental Protection Agency. Retrieved 27, August 2020

Permissions

All chapters in this book are published with permission under the Creative Commons Attribution Share Alike License or equivalent. Every chapter published in this book has been scrutinized by our experts. Their significance has been extensively debated. The topics covered herein carry significant information for a comprehensive understanding. They may even be implemented as practical applications or may be referred to as a beginning point for further studies.

We would like to thank the editorial team for lending their expertise to make the book truly unique. They have played a crucial role in the development of this book. Without their invaluable contributions this book wouldn't have been possible. They have made vital efforts to compile up to date information on the varied aspects of this subject to make this book a valuable addition to the collection of many professionals and students.

This book was conceptualized with the vision of imparting up-to-date and integrated information in this field. To ensure the same, a matchless editorial board was set up. Every individual on the board went through rigorous rounds of assessment to prove their worth. After which they invested a large part of their time researching and compiling the most relevant data for our readers.

The editorial board has been involved in producing this book since its inception. They have spent rigorous hours researching and exploring the diverse topics which have resulted in the successful publishing of this book. They have passed on their knowledge of decades through this book. To expedite this challenging task, the publisher supported the team at every step. A small team of assistant editors was also appointed to further simplify the editing procedure and attain best results for the readers.

Apart from the editorial board, the designing team has also invested a significant amount of their time in understanding the subject and creating the most relevant covers. They scrutinized every image to scout for the most suitable representation of the subject and create an appropriate cover for the book.

The publishing team has been an ardent support to the editorial, designing and production team. Their endless efforts to recruit the best for this project, has resulted in the accomplishment of this book. They are a veteran in the field of academics and their pool of knowledge is as vast as their experience in printing. Their expertise and guidance has proved useful at every step. Their uncompromising quality standards have made this book an exceptional effort. Their encouragement from time to time has been an inspiration for everyone.

The publisher and the editorial board hope that this book will prove to be a valuable piece of knowledge for students, practitioners and scholars across the globe.

Index

www.ingramcontent.com/pod-product-compliance
Lightning Source LLC
LaVergne TN
LVHW080717210926
848485LV00003B/478